Lecture Notes in Mechanical Engineering

Series Editors

Francisco Cavas-Martínez⊙, Departamento de Estructuras, Universidad Politécnica de Cartagena, Cartagena, Murcia, Spain

Fakher Chaari, National School of Engineers, University of Sfax, Sfax, Tunisia

Francesca di Mare, Institute of Energy Technology, Ruhr-Universität Bochum, Bochum, Nordrhein-Westfalen, Germany

Francesco Gherardini⊙, Dipartimento di Ingegneria, Università di Modena e Reggio Emilia, Modena, Italy

Mohamed Haddar, National School of Engineers of Sfax (ENIS), Sfax, Tunisia

Vitalii Ivanov, Department of Manufacturing Engineering, Machines and Tools, Sumy State University, Sumy, Ukraine

Young W. Kwon, Department of Manufacturing Engineering and Aerospace Engineering, Graduate School of Engineering and Applied Science, Monterey, CA, USA

Justyna Trojanowska, Poznan University of Technology, Poznan, Poland

Lecture Notes in Mechanical Engineering (LNME) publishes the latest developments in Mechanical Engineering—quickly, informally and with high quality. Original research reported in proceedings and post-proceedings represents the core of LNME. Volumes published in LNME embrace all aspects, subfields and new challenges of mechanical engineering. Topics in the series include:

- Engineering Design
- Machinery and Machine Elements
- Mechanical Structures and Stress Analysis
- Automotive Engineering
- Engine Technology
- Aerospace Technology and Astronautics
- Nanotechnology and Microengineering
- Control, Robotics, Mechatronics
- MEMS
- Theoretical and Applied Mechanics
- Dynamical Systems, Control
- Fluid Mechanics
- Engineering Thermodynamics, Heat and Mass Transfer
- Manufacturing
- Precision Engineering, Instrumentation, Measurement
- Materials Engineering
- Tribology and Surface Technology

To submit a proposal or request further information, please contact the Springer Editor of your location:

China: Ms. Ella Zhang at ella.zhang@springer.com
India: Priya Vyas at priya.vyas@springer.com
Rest of Asia, Australia, New Zealand: Swati Meherishi
at swati.meherishi@springer.com
All other countries: Dr. Leontina Di Cecco at Leontina.dicecco@springer.com

To submit a proposal for a monograph, please check our Springer Tracts in Mechanical Engineering at https://link.springer.com/bookseries/11693 or contact Leontina.dicecco@springer.com

Indexed by SCOPUS. All books published in the series are submitted for consideration in Web of Science.

More information about this series at https://link.springer.com/bookseries/11236

Rallapalli Srinivas · Rajesh Kumar · Mainak Dutta
Editors

Advances in Computational Modeling and Simulation

 Springer

Editors
Rallapalli Srinivas
Department of Civil Engineering
Birla Institute of Technology and Science
Pilani, Rajasthan, India

Rajesh Kumar
Department of Mathematics
Birla Institute of Technology and Science
Pilani, Rajasthan, India

Mainak Dutta
Department of Biotechnology
Birla Institute of Technology and Science
Dubai, United Arab Emirates

ISSN 2195-4356 ISSN 2195-4364 (electronic)
Lecture Notes in Mechanical Engineering
ISBN 978-981-16-7859-2 ISBN 978-981-16-7857-8 (eBook)
https://doi.org/10.1007/978-981-16-7857-8

This Springer imprint is published by the registered company Springer Nature Singapore Pte Ltd.
The registered company address is: 152 Beach Road, #21-01/04 Gateway East, Singapore 189721,
Singapore

Contents

About the Editors

Dr. Rallapalli Srinivas is presently working as an assistant professor in the Department of Civil Engineering, BITS Pilani-Pilani Campus, Rajasthan (India). His research interests are watershed modeling and simulation, environmental hydrology, hydro-climatology, and fuzzy logic-based decision making. Before joining BITS, he worked at University of Minnesota (UMN) as a Postdoctoral researcher and course instructor, U.S.A and successfully completed a project namely developing a field-scale optimized watershed modeling framework. He obtained his M.E (Civil Engineering with specialization in Infrastructure Systems) from Birla Institute of Technology and Science, Pilani in 2014, and B.E Hons. (Civil Engineering) and M.Sc. Hons. (Mathematics) from the same Institute in 2012. He was nominated by UMN Water Resources Center to represent the university at Michigan Data Science Symposium in 2019. He has also been invited to deliver keynote lectures at University of Michigan, University of Wisconsin, Iowa State University, National Resources Conservation Services (USA), Minnesota Association of Watershed Districts (USA) and Incessant Ganga Conference, Bihar Government (India). He has published over 30 papers in journals and conferences of international repute. Dr. Srinivas is a member of Soil and Water Conservation Society (USA), American Society of Civil Engineers, and the Institution of Engineers. For his research and teaching, he is actively collaborating with University of Minnesota, University of Wisconsin, Iowa State University and Minnesota Board of Water and Soil Resources. He is also a reviewer of leading journals such as Journal of Hydrology, Science of the Total Environment, Water Environment Research and Environmental Development and Sustainability. He has also done significant work on water quality modeling and simulation of River Ganga during his Ph.D. titled as 'Multi-criteria decision analysis and modeling for water quality management in Ganga river basin'.

Dr. Rajesh Kumar is currently an assistant professor at Department of Mathematics, BITS Pilani, Pilani Campus, Rajasthan, India. He obtained his M.Sc. in Applied Mathematics from IIT Roorkee. Further, he was selected for prestigious **Erasmus Mundus** fellowship and received Double Degree M.Sc. in "Industrial Mathematics: Modelling and Scientific Computing" from Technical University of Kaiserslautern,

Germany and Johannes Kepler University of Linz, Austria. Then, he finished his Ph.D. from Otto-von-Guericke University Magdeburg, Germany followed by Scientific Collaborator positions at MOX Milano Italy, EPFL Switzerland and continued further to work as a Research Scientist at Johann Radon Institute for Computational and Applied Mathematics (RICAM) Austria. During his educational tenure, he was recipient of **Deutsche Forschungsgemeinschaft (DFG) fellowship** by German Research Council and also got financial support by **Austrian academy of sciences** just to name a few. His major areas of research interests include "Mathematical and Numerical Analysis of Integro-Partial Differential Equations, Finite Volume Schemes, Low Rank Tensor Approximations and Uncertainty Quantifications". He has published 15 papers in respected international journals and organized several national and international workshops in the area of his research expertise.

Dr. Mainak Dutta is currently an assistant professor at the Department of Biotechnology, Birla Institute of Technology and Science, Pilani - Dubai. He was awarded Ph.D. from the Indian Institute of Technology Kharagpur, India. Dr. Dutta is the recipient of Fulbright USA Doctoral Research Fellowship, United States-India Educational Foundation (USIEF), 2014 and Sigma Xi Grants-in-Aid for graduate research, Sigma Scientific Research Society, USA, 2015. He has published more than 23 research articles in reputed peer-reviewed journals. His research interests lie in the use of proteomics, metabolomics and lipidomics approaches to study human diseases. In addition, he is also interested in computational approaches to drug design.

Adsorption of Ammonia on p-doped Graphene Bilayer Surface; Energetics and Electronic Structure

A. Sahithi and K. Sumithra

Abstract The adsorption of ammonia on pristine AB stacked graphene bilayers and also on p-doped surfaces of bilayer graphene are investigated using first principles density functional calculations. Modifications of the adsorption interactions and electronic structure effects due to doping and adsorption of ammonia are discussed. The adsorption of NH_3 is investigated for different dopant concentrations and for varied configurational patterns on the bilayer. Some of the bilayer configurations have strong interactions with ammonia depending on the dopant pattern and is evidenced by appreciable binding energies and charge transfer. The chemisorptions are confirmed by strong mixing of the non-bonding p orbitals of ammonia and with the electron deficient p-orbitals of the surface. The theoretical results on adsorption energies on doped bilayer are higher compared to adsorption on doped monolayer graphene more than 0.65 eV. The boron doped graphene bilayer induce the donor state above the Fermi level making it useful for sensing ammonia gas. The changes in the electronic properties of the system due to the interactions are expected to give useful understanding into the development of novel gas sensor devices.

1 Introduction

Graphene, the thinnest material known to man, has received much attention from a wide variety of fields, ever since its discovery. The physical properties of this material have been widely studied yielding various unusual phenomena, making it useful in super capacitors, field effect transistors, molecular storage etc. [1–3]. The high charge carrier mobility as well as the presence of Dirac points in the electronic structure makes it a gapless semiconductor [4–6]. Several studies have been performed on graphene in an attempt to change the carrier concentration and open a tuneable band gap. A myriad of ways was found to control the band gap which include doping with

A. Sahithi · K. Sumithra (✉)
Department of Chemistry, Birla Institute of Technology and Science (BITS), Pilani, Hyderabad Campus, Shamirpet, TG 500078, India
e-mail: sumithra@hyderabad.bits-pilani.ac.in

© The Author(s), under exclusive license to Springer Nature Singapore Pte Ltd. 2022
R. Srinivas et al. (eds.), *Advances in Computational Modeling and Simulation*,
Lecture Notes in Mechanical Engineering,
https://doi.org/10.1007/978-981-16-7857-8_1

impurities, applying an electric field, inducing defects and strain engineering [7, 8]. Multi-layered graphene, and in particular, bilayer graphene, has also attracted great spotlight as, surprisingly, it shows different electronic properties from monolayer graphene, while retaining much of the physical features of the monolayer [9]. It has also been observed that multi-layered graphene shows unique properties at room temperature such as ballistic transport [10].

Bilayer Graphene (BLG) is a two layered structure where a graphene layer is stacked on top of another graphene layer in two different stackings, namely AA and AB or Bernal stacking. Bernal stacking has been experimentally observed to be the preferred stacking arrangement for Bilayers. Electronic structures of both AA and AB stacked BLGs were studied by Schwingenschlögl et al. [9] and Saito et al. [11]. These density functional studies [9] on bilayer graphene showed that the interlayer distance after optimisation for AA is 3.6 Å and for AB is 3.4 Å. In AA stacking, the Dirac point is split into two points slightly away from the K point and in AB stacking, an overlap of the valence band and conduction band is seen at the K point, with the dispersion around the K point being parabolic as opposed to linear in monolayer graphene. Interestingly, BLG show semi-metallic nature in both the AB and AA stacking.

The effects of dopants on the structures and electronic properties of BLG have also been studied [9, 11]. The electronic band structures indicate the boron-doped BLG with characteristics of *p-doping* and nitrogen-doped BLG with *n-doping*, with a band gap opening. Saito et al. [11] also studied the effect of dopant position in the AB stacking, substituting the dopant atom in each of the non-equivalent sites on the upper layer. All these studies demonstrate doping of BLG as an effective way, not only to open the band gaps, but also to tune the magnitude of the band gaps. Furthermore, it has been previously reported that graphene is sensitive to the adsorption of gas molecules and many reports exist on the application of graphene and doped-graphene based systems for the developments of sensors and other nano-electronic devices [12–14]. The bilayer and multi-layered graphene have not been under much of this rapid development till recent times except for few studies.

2   Adsorption of Small Gas Molecules on Un-Doped and Doped BLG

There has been a huge interest in the recent years on adsorption and sensing of small gaseous molecules, especially atmospheric pollutants, on un-doped and doped bilayer graphene, owing to the variety of applications that they may offer [15–22]. It is observed in a Density Functional Theory (DFT) study of the effect of F_2 adsorption on the properties of Bilayer Graphene, by Shayeganfar [15], that the geometry of the F_2 adsorption could tune the induced band gap in BLG. Henrard et al. [16] investigated the binding energies and electronic properties of F_2 adsorption on the upper single layer of BLG. It is found that the binding energy of adsorption of F atoms depend on the sub lattice of the carbon atoms bonded with the fluorine atoms. Briddon et al. [17] studied the effect of water and ammonia molecules on bilayer graphene using DFT with Local Density Approximations. The binding energies of both ammonia and water on one of the layers of the bilayer graphene are found to be lesser than that of monolayer graphene where the decrease in BE is attributed to the sharing of electrons to the bottom layer. An electronic device using BLG has been fabricated [18], to experimentally study the sensing properties of BLG for CO_2 gas at room temperature. The electronic transport in BLG was strongly affected by the physisorption of CO_2 gas molecules onto the surface.

Multiple competing stable configurations that depend on the arrangement of the hydrogen atoms were found in the study of adsorption of hydrogen adatoms on bilayer graphene is observed in a studied by Chetty et al. [19]. The interaction of NO_2 molecules with bilayer graphene, both free-standing and on a SiC (0001) substrate was examined by Abrikosov et al. [20]. The binding energy values with and without substrate being considered were similar, with the free-standing bilayer having a higher BE of -114 meV. Band structures of the free-standing BLG after NO_2 adsorption showed *p-doping* character, with the Dirac point shifting above the Fermi level by 0.18 eV. It is seen that the substrate plays an important role in the properties of the system and must be taken into consideration for applications involving sensing of gases using BLGs. There have also been a few investigations of adsorption on doped BLG. Saito et al. [21] employed a DFT approach to study the effects of *p*- and *n*- *type* doping of bilayer, on adsorption of toxic gases (CO, CO_2, NO, NO_2). Low binding energies show that all the four gas molecules are physically adsorbed on nitrogen doped BLG. For boron doped bilayer, CO and CO_2 are physisorbed whereas NO and NO_2 molecules are chemically bound to the boron atom in the bilayer. Dai et al. [22] studied the performances of Fe doped bilayer graphene to study the adsorption gases such as NO, CO, HCN and SO_2. All the gas molecules were chemically adsorbed onto the BLG with a charge transfer from the Fe atom to the gas molecule.

The focus in the current article is on the adsorption of ammonia molecule onto undoped and boron doped bilayer graphene systems and the subsequent effect of the interactions on the measurable properties of the system. The effects of external parameters such as substrates and doping on the interactions are covered. In this article, we consider the adsorption of ammonia molecule on graphene bilayer surfaces doped with different mole fractions having different patterns.

3   Computational Methodology

All the density functional theory (DFT) calculations are done using VASP [23] with the projector augmented wave (PAW) [24] basis sets and periodic boundary conditions [25]. The generalized gradient approximation (GGA) [26] with the Perdew-Burke-Ernzerhof (PBE) [27] exchange–correlation functional is used, and the plane-wave cut-off energy is set as 520 eV in the calculations. The simulated system consists of a 4 × 4 graphene supercell bilayer with 64 carbon atoms with a doped atom substituting a carbon atom in the top layer and a single NH_3 molecule above the layer. van der Waals correction of DFT-D3 BJ damping [28] is used. The dopant concentrations that are considered are 1.56%. 3.12, 4.68%. The supercell extended for more than 18 Å in the direction normal to the graphene surface, in order to avoid the intervention between the images. In the geometrical structure optimization and self-consistent calculation, the Brillouin zone is sampled using a $12 \times 12 \times 1$ Monkhorst–Pack k-point grid [29] and Methfessel-Paxton smearing of 0.2 eV, which is tested to give converged results for all the properties calculated. Atomic positions are optimized until the maximum force on any ion is less than 0.02 eV/Å for all systems.

The binding or the adsorption energy is the most important physical quantity to reflect the adsorption strength and is calculated following Eq. (1):

$$E_{ad} = E_{gas+BLG} - E_{BLG} - E_{gas}, \tag{1}$$

where $E_{gas+BLG}$, E_{BLG} and E_{gas} are the total energies of the adsorbed system, isolated BLG and the gas molecule, respectively. The negative values of the energy imply affinity to the surface and vice versa. Charge transfer Δ_ρ between NH_3 and the bilayer system is calculated following Bader charge analysis [30]. For the adsorbed systems, Δ_ρ can be calculated as the charge variation of gas molecules before and after the adsorption, as below,

$$\Delta\rho = \rho(gas\,molecule + surface) - (\rho(surface) + \rho(gas\,molecule)). \tag{2}$$

4 Results and Discussions

We have considered adsorption on doped AB stacked bilayer graphene surface, where different doping patterns and mole-fractions are investigated. AB stacking is preferred for all calculations due to the stability over the AA stacking. In order to check the validity of the simulations, a proper referencing is made with the previous results of un-doped bilayer and single boron atom doped bilayers and compared the results with the existing results [7, 9, 11]. Both the calculations show matching results with the previous studies [9, 11], and the calculated band structure for the single atom doped bilayer with the mole fraction 1.56% is comparable to that made by Schwingenschlögl et al. [9]. The results, the band structures and the corresponding density of states on un-doped BLG and singly doped BLG are given together in Fig. 1.

The interlayer distance for the intrinsic bilayer is found to be 3.4 Å for AB stacking, which is in accordance with the experimental results [31]. These values are modified by the presence of dopant atoms for the concentrations considered in this study, and an in-plane lattice parameter of ~ 2.5 Å is observed for all the doped surfaces. The

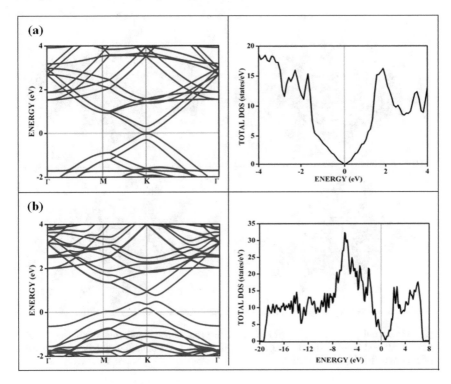

Fig. 1 The calculated band structure and the corresponding density of states of **a** intrinsic bilayer and **b** doped bilayer graphene in AB stacking with 1.56% boron depicted as 1B-AB

different configurations studied with two different mole fractions of doping 3.12 and 4.68% are shown in Fig. 2.

Two concealed Dirac cones can be seen at the K point in Fig. 1b, which can be attributed to the layers doped with B (unoccupied cone) and the bare graphene layer and a band gap of 0.28 eV is observed. Such buried cones are also visible for the other configurations in the band structures in Fig. 3, for the different surfaces mentioned in Fig. 1. The energetically favourable distance of ~ 2.5 Å indicates favourable π-π interactions between the layers. The electronic band structure of the minimum energy configurations shown in Fig. 3, indicates that the occupation of the various sub-lattice positions have no influence on the band gap observed, unlike doping in the case of single layered graphene [12]. The parabolic bands can be seen in the vicinity of the Fermi energy, as expected for AB stacked bilayer graphene for all the cases. The band gaps corresponding to these different configurations of BLG are given in Table 1.

It is seen that the band dispersions are not very much affected, except in the case of certain configurations, and doping in moderate concentrations lead to a band gap of about 350 meV, on which adsorptions are studied. The B-doped graphene systems possess holes in the valence bands and are *p-type* materials on which ammonia with

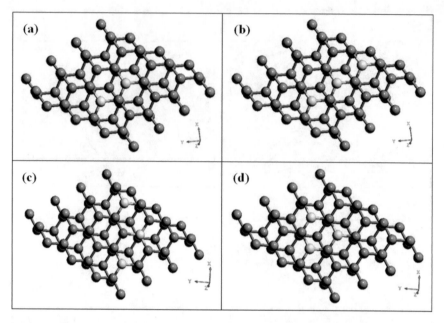

Fig. 2 Optimized geometries of doped bilayer structures, depicted with the notations **a** b-AB **b** d-AB **c** e-AB and **d** f-AB. b-AB corresponds to the dopant concentration of 3.12% and all other configurations (**b**)–(**d**) correspond to 4.68%

its lone pair of electrons adsorbs strongly. The results of the adsorption of ammonia on these surfaces are summarised in Table 2.

It is interesting to note that the adsorption energies of NH_3 on some of the boron-doped bilayer graphene, for example on the 'd-AB' and the 'e-AB' surface are strong and the adsorption energies and charge transfers correspond to that of typical chemisorption. On the other hand, on all other configurations, it is only physisorbed with negligible charge transfer and adsorption energies. The difference in the adsorptions can be explained in terms of charge differences occurring in various substrates. In the configurations where it is chemisorbed, boron atom are much more electron deficient than in the other patterns and therefore forms a bond with nitrogen of ammonia. The B–N distance in this case is of the order of 1.66Å, which falls in the typical range of B-N single bond distance. Similar adsorption has also been observed for ammonia adsorption on doped carbon nanotubes [32].

The electronic structure of these adsorption studies is summarised in Fig. 4 where the structures are shown alongside the electronic band diagram and the corresponding density of states. It is interesting to note that the electronic band gaps change by few meV for the cases of strong chemisorption while remaining unaltered for the physisorptions. Though these are not appreciable, these changes are expected to cause changes in transport properties. The adsorption energies are appreciable for the NH_3/d-AB and NH_3/e-AB and is of the order of -1 eV. The density of states of

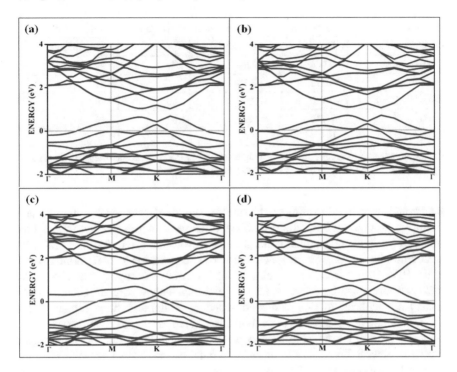

Fig. 3 Band structures of stable BLG structures **a** b-AB **b** d-AB **c** e-AB and **d** f-AB

Table 1 Band gap and the optimised interlayer distanced for graphene bilayer surfaces with dopant of different concentrations and patterns

System	Band gap (eV)	D (Å)
1B-AB	0.28	2.68
b-AB	0.34	2.65
d-AB	0.35	2.71
e-AB	0.38	2.71
f-AB	0.12	2.37

D is the interlayer distance after optimisation

ammonia shown in red indicates that there is a shift to lower energy for two of the bands. In order to understand the chemisorptive behaviour, this is further analysed.

Analysing the density of states corresponding to the case where chemisorption is observed in the case of surfaces d-AB and e-AB, orbital hybridization happens between the '2a1' orbital, which is a non-bonding p orbital on nitrogen of ammonia and the p orbital on boron. In Fig. 5, the density of states corresponding to ammonia

Table 2 Adsorption energies, favourable distances and the band gaps for the different systems

System	(^1E_a) (eV)	D^1 (Å)	D^2 (Å)	Band gap (eV)
NH$_3$/1B-AB	−0.046	2.59	3.2	0.26
NH$_3$/b-AB	−0.222	2.59	3.4	0.31
NH$_3$/d-AB	−0.989	2.9	1.66	0.28
NH$_3$/e-AB	−1.012	2.95	1.66	0.16
NH$_3$/f-AB	−0.284	2.62	3	0.12

^1E_a is the adsorption energy of bilayer graphenes; D^1 is the distance between two layers after optimisation; D^2 is the distance between adsorbate and adsorbent

before and after the adsorption on the surface type 'd-AB' is shown. Before adsorption, (represented by black line) it shows three peaks corresponding to the three molecular orbitals of ammonia and after the adsorption (blue line) the peak near the Fermi disappears. The peak near the Fermi corresponds to '2a1' orbital which is the HOMO, non-bonding p orbital of nitrogen and is occupied. The absence of this peak corresponding to 2a1 in (d) and (e) in Fig. 4 shows that the chemisorption takes place with the mixing of this non-bonding orbital and the p- orbitals of the electron deficient boron to form the bond. For comparison, the electronic structure corresponding to un-doped bilayer, doped bilayer and adsorption of ammonia are drawn together in Fig. 6. The changes in the band gap are visible from this plot, and there is loss of dispersion after the adsorption.

The software, VESTA (Visualization for electronic and structure analysis) [33] is used to extract and visualise useful information on the electronic structure and charge transfer contours from the first-principles calculations. The contour plots of the electronic charge density of the valence and conduction bands of NH$_3$/d-AB are shown in Fig. 7. To understand the bond formation of B-N bonds further, the electron density contour plots are constructed. The plane hkl (001) containing N-B-C bonds are used, for various partitions of electronic charge densities of NH$_3$/d-AB.

The charge density difference between the adsorbed complex, the surface and the gas molecule NH$_3$ is calculated using Eq. 2, and is plotted in Fig. 7 in two different outlooks (a) and (b), to see the changes due to adsorption. The electronic changes due to ammonia adsorption is clear from the difference density plot. The charge density reorganization takes place due to chemisorptive adsorption. It is possible to see electron deficient regions in the immediate proximity of boron atoms compared to the rest of the surface. The red colour from the figure represents nitrogen with high charge density, yellow iso-charge surfaces represent 2.24e-05 e/Å^3 of charge accumulation, and the bottom cyan colour shows 0.023 e/Å^3 charge depletion on the graphene surface. Moreover, due to the charge accumulated around the nitrogen, it is attracted towards the charge deficient region (cyan iso-surface, i.e. boron of graphene bilayer iso-surface) adding to the energetic preference of nitrogen to bond with the particular boron which is AB stacked to the bottom layer. Here, the valence charge (cyan) around boron atoms is attracted towards the nitrogen atom (from the cyan toward the yellow iso-charge surface), resulting in the formation of a covalent

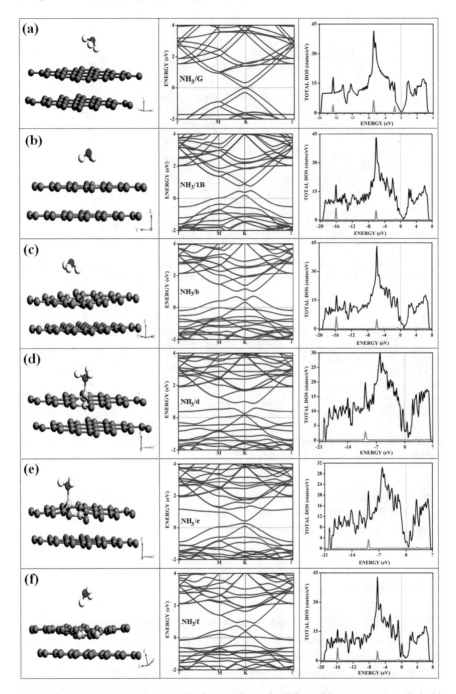

Fig. 4 The band structures and respective density of sates for different bilayer systems studied with ammonia as adsorbate **a** NH_3/G-AB (Ammonia on un-doped AB stacked BLG) **b** $NH_3/1B$-AB **c** NH_3/b-AB **d** NH_3/d-AB **e** NH_3/e-AB **f** NH_3/f-AB. The partial density of states for ammonia for these adsorption processes are shown in red

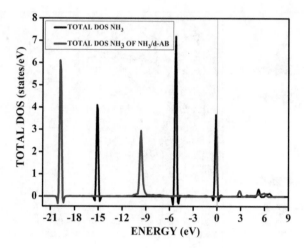

Fig. 5 Total DOS of ammonia before and after adsorption on surface d-AB of bilayer graphene. The absence of the blue peak near Fermi indicates the orbital hybridization

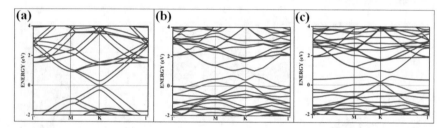

Fig. 6 Band Structures of geometry with doped monolayer, doped and with ammonia as adsorbate respectively **a** G-AB **b** d-AB **c** NH$_3$/d-AB

bond. The bonding states of B–N bonds are on top of the valence bands. The figure shows that the concentration of charge is distributed among the carbon atoms that are adjacent to borons or have formed bonds with borons.

5 Conclusion

Graphene bilayer and its doped derivatives clearly show favourable properties towards adsorption of gases, and correspondingly, show potential for the development of gas sensor devices. As has discussed in the article, the doping with moderate concentrations and adsorption of ammonia on BLG show specific changes in the electronic properties. The electronic structure for doped systems indicates a band gap opening for the semi-metallic bilayer and adsorption is favoured on some of the surface patterns. This could lead to unique dopant combinations for the sensing

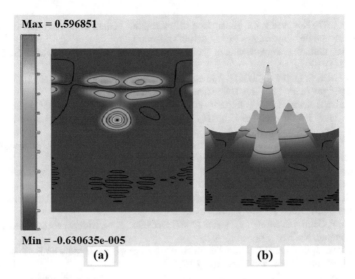

Fig. 7 Electron charge density contour map of ammonia adsorption on the surface d-AB (NH_3/d-AB) **a** contour plot and **b** the bird eye view perspective

of various specific gases. Controlling of adsorption parameters by use of dopant concentrations and varied configurations are observed and can be a key factor in the development of devices to improve the gas sensing performance.

Overall, the future of bilayer graphene and its derivatives in the field of gas adsorption and sensing seems promising and could also potentially extend to the storage and capture of gases. Much effort remains to be done, in understanding the effect of dopants towards the adsorption of various non-toxic gases such as NH_3, CH_4, H_2O, H_2 and O_2. Advancements for the practical applications of these chemically modified bilayers include its prospective, potential application in the sensing of pollutants like ammonia.

Acknowledgements Andru Sahithi thanks the support from Birla Institute of Technology and Science, Hyderabad, campus in the form of scholarship.

References

1. Novoselov KS, Geim AK, Morozov SV et al (2004) Electric field effect in atomically thin carbon films. Science 306(5696):666–669
2. Castro EV, Novoselov KS, Morozov SV et al (2007) Biased bilayer graphene: semiconductor with a gap tunable by the electric field effect. Phys Rev Lett 99(21):216802
3. Zhang Y, Tang TT, Girit C et al (2009) Direct observation of a widely tunable bandgap in bilayer graphene. Nature 459(7248):820–823
4. Bolotin KI, Sikes KJ, Jiang Z et al (2008) Ultrahigh electron mobility in suspended grapheme. Solid State Commun 146(9–10):351–355

5. Morozov SV, Novoselov KS, Katsnelson MI et al (2008) Giant intrinsic carrier mobilities in graphene and its bilayer. Phys Rev Lett 100(1):016602
6. Neto AC, Guinea F, Peres NM et al (2009) The electronic properties of graphene. Rev Mod Phys 81(1):109
7. Nanda BRK, Satpathy S (2009) Strain and electric field modulation of the electronic structure of bilayer graphene. Phys Rev B Condens Matter 80(16):165430
8. Altanhana T, Kozal B (2012) Impurity effects in graphene. Eur Phys J B 85(7):222
9. Alattas M, Schwingenschlögl U (2018) Band gap control in bilayer graphene by co-doping with B-N pairs. Sci Rep 8(1):1–6
10. Berger C, Song Z, Li X et al (2006) Electronic confinement and coherence in patterned epitaxial grapheme. Science 312(5777):1191–1196
11. Fujimoto Y, Saito S (2015) Electronic structures and stabilities of bilayer graphene doped with boron and nitrogen. Surf Sci 634:57–61
12. Sahithi A, Sumithra K (2020) Adsorption and sensing of CO and NH_3 on chemically modified grapheme surfaces. RSC Adv 10:42318–42326
13. Schedin F, Geim AK, Morozov SV et al (2007) Detection of individual gas molecules adsorbed on grapheme. Nat Mater 6(9):652–655
14. Dai J, Yuan J, Giannozzi P (2009) Gas adsorption on graphene doped with B, N, Al, and S: a theoretical study. Appl Phys Lett 95(23):232105
15. Shayeganfar F (2015) Energy gap tuning of graphene layers with single molecular F_2 adsorption. J Phys Chem C 119(22):12681–12689
16. Santos H, Henrard L (2014) Fluorine adsorption on single and bilayer graphene: role of sublattice and layer decoupling. J Phys Chem C 118(46):27074–27080
17. Ribeiro RM, Peres NMR, Coutinho J et al (2008) Inducing energy gaps in monolayer and bilayer graphene: local density approximation calculations. Phys Rev B 78(7):075442
18. Sun J, Muruganathan M, Mizuta H (2016) Room temperature detection of individual molecular physisorption using suspended bilayer grapheme. Sci Adv 2(4):e1501518
19. Mapasha RE, Ukpong AM, Chetty N (2012) Ab initio studies of hydrogen adatoms on bilayer grapheme. Phys Rev B 85(20):205402
20. Caffrey NM, Armiento R, Yakimova R et al (2016) Changes in work function due to NO_2 adsorption on monolayer and bilayer epitaxial graphene on SiC (0001). Phys Rev B 94(20):205411
21. Fujimoto Y, Saito S (2016) Gas adsorption, energetics and electronic properties of boron-and nitrogen-doped bilayer graphenes. Chem Phys 478:55–61
22. Tang Y, Liu Z, Shen Z et al (2017) Adsorption sensitivity of metal atom decorated bilayer graphene toward toxic gas molecules (CO, NO, SO_2 and HCN). Sens Actuators B Chem 238:182–195
23. Eyert V, Christensen M, Wolf W et al (2018) Unravelling the potential of density functional theory through integrated computational environments: recent applications of the Vienna Ab initio Simulation Package in the MedeA® Software, Computation 6
24. Kresse G, Joubert D (1999) From ultrasoft pseudopotentials to the projector augmented wave method. Phys Rev B: Conden Matter Mater 59:1758
25. Makov G, Payne MC (1995) Periodic boundary conditions in ab initio calculations. Phys Rev B 51:4014
26. Grimme S (2006) Semiempirical GGA-type density functional constructed with a long-range dispersion correction. J Comput Chem 27:1787
27. Perdew JP, Burke K, Ernzerhof M (1997) Generalized gradient approximation made simple. Phys Rev Lett 78:1396
28. Grimme S, Antony J, Ehrlich S et al (2010) A consistent and accurate *ab initio* parametrization of density functional dispersion correction (DFT-D) for the 94 elements H-Pu. J Chem Phys 132 (15):154104
29. Monkhorst HJ, Pack JD (1976) Special points for Brillouin-zone integrations*. Phys Rev B 13:5188

30. Henkelman G, Arnaldsson A, Jónsson H (2006) A fast and robust algorithm for Bader decomposition of charge density. Comput Mater Sci 36(3):354–360
31. Lee JK, Lee SC, Ahn JP et al (2008) The growth of AA graphite on (111) diamond. J Chem Phys 129(23):234709
32. Vikramaditya T, Sumithra K (2014) Effect of substitutionally boron-doped single-walled semiconducting zigzag carbon nanotubes on ammonia adsorption. J Comput Chem 35(7):586–594
33. Momma K, Izumi F (2011) VESTA 3 for three-dimensional visualization of crystal, volumetric and morphology data. J Appl Crystallog 44:1272–1276

Design and Analysis of Complex Computer Models

Jeevan Jankar, Hongzhi Wang, Lauren Rose Wilkes, Qian Xiao, and Abhyuday Mandal

Abstract This chapter presents a review of some state-of-the-art statistical techniques for analyzing real computer experiments which play a significant role in various scientific research and industrial applications. In computer experiments, emulators (i.e. surrogate models) are often used to rapidly approximate the outcomes and reduce the computational expense. Gaussian process (GP) models, also known as Kriging, are a common choice of emulators, and optimal experimental designs should be used to improve their accuracy. Specifically, space-filling designs are widely used in the literature, which proved to be efficient under GP models. In this chapter, we review different types of GP models as well as various kinds of space-filling designs. We further provide a practical tutorial on how to construct space-filling designs and fit GP emulators to analyze real computer experiments.

Keywords Computer experiments · Gaussian process models · Space-filling designs · Latin hypercube designs

1 Introduction

A computer experiment is a system of complex computer codes simulating a physical process. They are implemented like a function, taking inputs to produce the outputs. This automation can reduce the cost, time, and/or management compared to a traditional lab experiment (see, for example, [20]). Computer experiments are often deterministic (specified inputs will always produce the same output), making the results more stable and less prone to random errors compared to traditional lab experiments. Researchers can manipulate the code to systematically adjust a wide range of inputs and generate outputs based on what they are trying to study. They are instrumental in cases where a physical experiment would be impossible, such as modeling black holes [29]. Due to these characteristics, computer experiments

J. Jankar · H. Wang · L. R. Wilkes · Q. Xiao · A. Mandal (✉)
University of Georgia, Athens, GA 30602, Georgia
e-mail: abhyuday@uga.edu; amandal@stat.uga.edu

© The Author(s), under exclusive license to Springer Nature Singapore Pte Ltd. 2022
R. Srinivas et al. (eds.), *Advances in Computational Modeling and Simulation*,
Lecture Notes in Mechanical Engineering,
https://doi.org/10.1007/978-981-16-7857-8_2

15

become very popular in various scientific research and industrial applications (see, for more examples, [12, 20]). For example, [8] created a 3D mixed finite element model to study flexoelectric material. The Flexoelectric Effect is where strain gradients polarize electric fields. This process is complicated to study, especially in a practical context, so the finite element method is a numerical approach, i.e. computer experiment, used to study this effect. Mixed finite elements simplify this task further using an alternative way of handling higher order derivatives.

Computer experiments are often computationally intensive, though computing power has increased in recent years. To rapidly generate many outcomes and reduce the computational expenses, emulators (i.e. surrogate models) are needed which are often fitted with only a few data points. Emulators should also allow uncertainty quantification to measure how accurate the model is for predictions. If a good emulator is selected, it may be more useful than the underlying physical process as it eliminates noise. The Gaussian Process (GP) model is a widely used emulator [20, 43]. The GP assumes all observations following a multivariate normal distribution, which is characterized by a mean vector μ and a covariance matrix Σ. The GP model would interpolate the observations, which is desirable for computer experiments having deterministic outputs. It also allows for accurate uncertainty quantification. By specifying different types of covariance functions, researchers may further add prior knowledge about the shape of the response surface.

The GP model has been applied to many computer experiments in Chemistry, Computational Biology, Robotics and others [30]. As an illustration, it has accurately simulated the collision dynamics of complex molecules [6], the spread of COVID-19 [52], flagging suspicious Internet claims [63] and autonomous learning in robots [7]. Data scientists at Microsoft introduced a framework that enables the application of GP models to data sets containing millions of data points [23]. As pictured in Fig. 1, a Bayesian framework is used for human body pose tracking [10]. Instead, a GP experiment can be used to take in a description of a human silhouette as inputs and outputs to identify human pose [68]. One useful application of GP in Astronomy is modeling the collision of two black holes. Researchers cannot create black holes to observe and experiment with, so computer experiments offer a veritable way to simulate the outcome of black hole collisions. Figure 2 illustrates

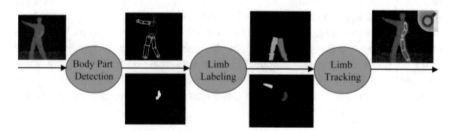

Fig. 1 An example of Bayesian framework for human pose tracking *Source* https://www.ncbi.nlm.nih.gov/pmc/articles/PMC3292173/ [68]

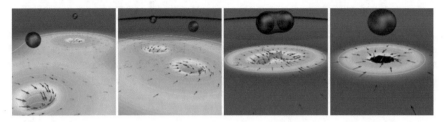

Fig. 2 Computer simulation of two black holes colliding *Source* https://www.black-holes.org/code/SpEC.html

that computer models and GP emulators are created based on the known properties of black holes and the surrounding system of space and are compared to naturally observed black hole movement in order to test how accurate they are [58]. Another interesting application of GP is on car crash simulation to study the damage on the car. Here, models are validated by comparing simulation results with an actually controlled crash. Figure 3 depicts some results from a finite element method.

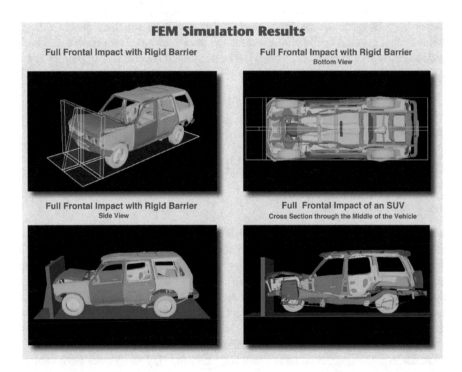

Fig. 3 An example of Gaussian Process experiment in car crash simulation *Source* https://www.csm.ornl.gov/SC98/car.html

The remainder of this chapter is organized as follows. In Sect. 2, we systematically review the GP models. Specifically, we discuss the ordinary and universal GP in Sect. 2.1, their model estimations and uncertainty quantification in Sect. 2.2 and methods for including qualitative inputs in Sect. 2.3. In Sect. 3, we review popular experimental designs used in computer experiments, and we conclude this chapter in Sect. 4.

2 The Gaussian Process Model

In this section, we aim to understand GP as a flexible nonparametric regression for surrogate modeling in computer experiments. GP is widely used in many statistical and probabilistic modeling enterprises. GP is a very generic term, and all it means is that any finite collection of realizations is modeled as having a multivariate normal (MVN) distribution. That means, a finite collection of n observations can be completely characterized by their mean vector μ and covariance matrix Σ.

Let $y(\mathbf{x_i})$ be the output which is assumed to be a deterministic real-valued function of the d-dimensional variable $\mathbf{x_i} = (x_{i1}, \ldots, x_{id})^T \in D \subset \Re^d$, for $i = 1, 2, \ldots, n$. Let $(Y_x)_{x \in D}$ be a square-integrable random field and y be a realization of $(Y_x)_{x \in D}$. Let $\mathbf{X} = \{\mathbf{x_1}, \ldots, \mathbf{x_n}\}$ be the points where their responses have been observed, which is denoted by $\mathbf{y} = (y(\mathbf{x_1}), \ldots, y(\mathbf{x_n}))^T$. The aim of GP is to optimally predict Y_x by a linear combination of the observations $\mathbf{y}$, for any $\mathbf{x} \in D$.

2.1 Model Formulation

Ordinary GP, also known as ordinary Kriging, has the form

$$y(\mathbf{x_i}) = \mu + Z(\mathbf{x_i}), \tag{1}$$

where μ is the mean vector and $Z(\mathbf{x_i})$ is a GP such that $Z(\mathbf{x_i}) \sim GP(0, \sigma^2\Sigma)$. In the above model, $Z(\mathbf{x_i})$ is GP with zero mean, and the covariance function $\phi(\cdot) = \sigma^2\Sigma(\cdot|\theta)$, where $\theta = (\theta_1, \ldots, \theta_d)^T$ is the vector of unknown correlation parameters with all $\theta_k > 0$ $(k = 1, \ldots, d)$ and Σ is a stationary correlation function that determines the correlation between inputs with parameters θ. The mean of the GP controls the trend, whereas the correlation function controls the smoothness of its sample paths. Power-exponential, Gaussian and Matérn correlation functions are the most widely used ones in the literature.

In the power-exponential correlation structure, the (i, j)th element in the correlation matrix is defined as follows:

$$\Sigma\left(\mathbf{x}_i, \mathbf{x}_j \mid \boldsymbol{\theta}\right) = \prod_{k=1}^{d} \exp\left\{-\theta_k \left|x_{ik} - x_{jk}\right|^{p_k}\right\} \quad \text{for all } i, j, \tag{2}$$

with two inputs $\mathbf{x}_i = (x_{i1}, \ldots, x_{id})^T$ and $\mathbf{x}_j = (x_{j1}, \ldots, x_{jd})^T$ and smoothness parameters $p_1, \ldots, p_d$, which lie between 0 and 2, with 0 giving the most rough results and 2 giving the most smooth. If we take $p_k = 2$ for all $k = 1, \ldots, d$, then it results in the popular Gaussian correlation function:

$$\Sigma\left(\mathbf{x}_i, \mathbf{x}_j \mid \boldsymbol{\theta}\right) = \prod_{k=1}^{d} \exp\left\{-\theta_k \left|x_{ik} - x_{jk}\right|^{2}\right\} \quad \text{for all } i, j. \tag{3}$$

The correlation functions of Matérn family is given by

$$\Sigma(\mathbf{h} \mid \boldsymbol{\theta}) = \prod_{k=1}^{d} \frac{1}{\Gamma(v)2^{v-1}} \left(\frac{2\sqrt{v}\,|h_k|}{\theta_k}\right)^{v} K_v\left(\frac{2\sqrt{v}\,|h_k|}{\theta_k}\right), \tag{4}$$

where $v > 0$ is a smoothness parameter, $\Gamma(\cdot)$ is the Gamma function and $K_v(\cdot)$ is the modified Bessel function of order v. Two commonly used orders are $v = 3/2$ and $v = 5/2$.

Different correlation functions mentioned above impose different characteristics for function draws, allowing for different properties when modeling computer models. For example, when using the power-exponential function, all sample paths are infinitely differentiable when $p_k = 2$. For the Matérn correlation function, when we have $d = 1$, all sample paths are $\lceil v \rceil - 1$ differentiable. Hence, v is viewed as a smoothness parameter.

In the literature, two important assumptions are often imposed on the ordinary GP model to effectively analyze computer experiment. One assumption is that the GP is separable [9], which means finite-dimensional distributions can determine sample path properties of function draws which are usually infinite-dimensional. The second important assumption is that the model is stationary. Consider $\{\mathbf{x}_1, \ldots, \mathbf{x}_n\} \in D$ and any $h \in \mathfrak{R}^d$, then a GP model is said to be stationary if the random vectors $(Y(\mathbf{x}_1), \ldots, Y(\mathbf{x}_n))$ and $(Y(\mathbf{x}_1 + \mathbf{h}), \ldots, Y(\mathbf{x}_n + \mathbf{h}))$ follow the same distribution. This means that both these random vectors should have the same mean and covariance.

The second assumption is restrictive, and we may need more flexibility while modeling computer experiments. One popular approach is to extend the above ordinary GP model to incorporate a global trend function for the mean part. This extended model is known as *Universal Kriging* which has the form:

$$y(\mathbf{x}) = \mu(\mathbf{x}) + Z(\mathbf{x}), \tag{5}$$

with $\mu(\mathbf{x}) = \mathbf{f}(\mathbf{x})^T \boldsymbol{\beta} = \sum_{s=1}^{m} \beta_s f_s(\mathbf{x})$, where $\mathbf{f}$ is a m-dimensional known function and $\boldsymbol{\beta} = (\beta_1, \ldots, \beta_m)^T$ is a vector of unknown parameters. The idea is to rely on functions in $\mathbf{f}(\mathbf{x})$ to de-trend the process and then model any residual variation as zero mean stationary GP. Taking constant mean $\mathbf{f}(\mathbf{x}) \equiv 1$ results in the ordinary GP model discussed above. The stationary correlation functions discussed above in Eqs. (2) and (4) can also be applied here, that is,

$$Cov\,(Z(\mathbf{x} + \mathbf{h}),\, Z(\mathbf{x})) = \sigma^2 \Sigma(\mathbf{h}),$$

where correlation function $\Sigma(\mathbf{h})$ is a positive semidefinite function with $\Sigma(\mathbf{0}) = 1$ and $\Sigma(\mathbf{h}) = \Sigma(-\mathbf{h})$.

2.2　Estimation and Uncertainty Quantification

In this section, we present equations used for predicting and quantifying uncertainty on $y(\mathbf{x})$ given observed responses $\mathbf{y} = (y(\mathbf{x_1}), \ldots, y(\mathbf{x_n}))^T$. The question we are trying to answer is: given examples of function in pairs $(\mathbf{x_1}, y(\mathbf{x_1})), \ldots, (\mathbf{x_n}, y(\mathbf{x_n}))$, what random function realizations could explain or could have generated those observed values? In other words, we want to calculate the conditional distribution $(Y(\mathbf{x_1}), \ldots, Y(\mathbf{x_n})) \,|\, \{(\mathbf{x_1}, y(\mathbf{x_1})), \ldots, (\mathbf{x_n}, y(\mathbf{x_n}))\}$.

Before we calculate the *predictive distribution*, we need to address the key question of how the parameters β, σ^2 and $\boldsymbol{\theta}$ are estimated from the data $(x_i, y(x_i))_{i=1}^n$. The most popular approach for parameter estimation is *maximum likelihood estimation*, and the log-likelihood function under the above assumed GP model can be written as

$$l\left(\beta, \sigma^2, \boldsymbol{\theta}\right) = -\frac{1}{2}\left[n \log \sigma^2 + \log \det \Sigma_\theta + \frac{1}{\sigma^2}\,(\mathbf{y} - \mathbf{F}\beta)\,\Sigma_\theta^{-1}\,(\mathbf{y} - \mathbf{F}\beta)\right], \quad (6)$$

where $\det \Sigma_\theta$ is the determinant of the matrix $\Sigma_\theta = \left[\Sigma(\mathbf{x_i}, \mathbf{x_j})\right]_{i=1}^{n}{}_{j=1}^{n}$ and $\mathbf{F} = [f_s(\mathbf{x_i})]_{i=1}^{n}{}_{s=1}^{m}$. Hence, the MLEs for $(\beta, \sigma^2, \boldsymbol{\theta})$ are the parameter estimates that maximize the above log-likelihood function. ML estimates of (β, σ^2) for fixed value of $\boldsymbol{\theta}$ can be easily obtained as follows:

$$\hat{\beta}_\theta = \left(\mathbf{F}^T \Sigma_\theta^{-1} \mathbf{F}\right)^{-1} \mathbf{F}^T \Sigma_\theta^{-1} \mathbf{y} \quad (7)$$

and

$$\hat{\sigma}_\theta^2 = \frac{1}{n}\left(\mathbf{y} - \mathbf{F}\hat{\beta}_\theta\right)^T \Sigma_\theta^{-1} \left(\mathbf{y} - \mathbf{F}\hat{\beta}_\theta\right). \quad (8)$$

Substituting these ML estimates back into Eq. (6), we get the profile likelihood function as follows:

$$l\left(\hat{\beta}, \hat{\sigma}^2, \boldsymbol{\theta}\right) = -\frac{1}{2}\left[n \log \hat{\sigma}^2 + \log \det \Sigma_\theta + n\right], \tag{9}$$

where the MLE of $\boldsymbol{\theta}$ is one that maximizes the above function in Eq. (9). This optimization problem does not enjoy a closed-form solution, so numerical methods, e.g. quasi-Newton algorithms [40] are used for solving the problem.

Once we have estimates of parameters, we can calculate the conditional distribution as mentioned above. Let $\left(\hat{\beta}, \hat{\sigma}^2, \hat{\boldsymbol{\theta}}\right)$ denote the ML estimates of unknown parameters for the given GP model. Then for a new input $\mathbf{x}^* \in \Re^d$, the mean and variance of random variable $Y(\mathbf{x}^*|\mathbf{y})$ are as follows:

$$\hat{y}\left(\mathbf{x}^*\right) = \mathbb{E}\left[Y\left(\mathbf{x}^*\right) \mid \mathbf{y}\right] = \mathbf{f}^T\left(\mathbf{x}^*\right)\hat{\beta} + \mathbf{r}_{\hat{\theta}}^T\left(\mathbf{x}^*\right)\Sigma_{\hat{\theta}}^{-1}\left(\mathbf{y} - \mathbf{F}\hat{\beta}\right), \tag{10}$$

$$s\left(\mathbf{x}^*\right)^2 = \text{Var}\left[Y\left(\mathbf{x}^*\right) \mid \mathbf{y}\right] = \hat{\sigma}^2\left(1 - \mathbf{r}_{\hat{\theta}}^T\left(\mathbf{x}^*\right)\Sigma_{\hat{\theta}}^{-1}\mathbf{r}_{\hat{\theta}}\left(\mathbf{x}^*\right)\right), \tag{11}$$

where the covariance vector $\mathbf{r}_{\hat{\theta}}\left(\mathbf{x}^*\right) = \left[\Sigma_{\hat{\theta}}\left(\mathbf{x}^*, \mathbf{x}_1\right), \Sigma_{\hat{\theta}}\left(\mathbf{x}^*, \mathbf{x}_2\right), \ldots, \Sigma_{\hat{\theta}}\left(\mathbf{x}^*, \mathbf{x}_n\right)\right]^T$.

When some observed data points are very close to each other, the covariance matrix $\Sigma_{\hat{\theta}}$ may become nearly singular, making it difficult to obtain a stable inverse matrix $\Sigma_{\hat{\theta}}^{-1}$. This is a common issue for GP models, when the run and/or factor sizes are large. One way to deal with this problem is to add a positive scalar λ, called the *nugget* parameter, to the diagonal elements in $\Sigma_{\hat{\theta}}$, i.e. replacing Σ_θ with $\Sigma_\theta + \lambda\mathbf{I}$, where $\mathbf{I}$ is an identity matrix. Adding λ is analogous to adding the ridge parameter in ridge regression, which helps in moving the smallest eigenvalue of Σ_θ away from zero, thus stabilizing the calculation of its inverse.

For large data sizes, the estimation of GP models can be very time-consuming, mainly due to the matrix inverse calculation of order $O(n^3)$. To deal with this problem, [21] proposed a localize GP (LaGP) approach. Based on a local subset of the data, they provide a family of local sequential design schemes that defines the support points of a GP predictor. The idea is to make sure that for a given choice of covariance structure, the data points far from the target location $\mathbf{x}^*$ will have little effect on the prediction. Hence, it is not unwise to calculate the inverse of the full covariance matrix, as the elements corresponding to "far away" points will contribute very little to predicting $y(\mathbf{x}^*)$. Interested readers may refer to [21] for further details.

The notion of calibration and sensitivity analysis is important in the context of physical and computer experiments. In practice, we only observe response y_{Field} instead of observing real physical response y_{Real}. And, we use the above computer models to approximate y_{Real} as y_{Model}. Now, as we saw in the earlier sections apart from input variables, computer models also use some more parameters known as calibration parameters to fine-tune the model. Covariance parameters $\boldsymbol{\theta}$ are one such example of calibration parameters. A Bayesian framework was proposed by [28] to address this as follows:

$$y_{Real}(\mathbf{x}) = y_{Model}(\mathbf{x}, \boldsymbol{\theta}) + \text{b}(\mathbf{x})$$
$$y_{Field}(\mathbf{x}) = y_{Model}(\mathbf{x}, \boldsymbol{\theta}) + \text{b}(\mathbf{x}) + \epsilon,$$

where $\text{b}(\mathbf{x})$ is a bias and ϵ is the normal error. Reference [28] used Bayesian methods to estimate the bias correction function and unknown calibration parameter $\boldsymbol{\theta}$ under a GP prior. Iterative history matching algorithm as one proposed by [53] for calibrating a galaxy formation model called GALFORM is an alternative to this Bayesian approach. Recently, [1] used this algorithm for calibrating hydrological time-series models.

2.3 GP with Qualitative Inputs

The above-mentioned GP model is valid only with quantitative inputs, but there are many situations in real life where inputs can be both quantitative and qualitative. One straightforward way to adapt GP models with qualitative inputs is to construct separate GP models for each level combination of the qualitative factors. Yet, when there are many high-level qualitative factors, such an approach would require many observations to fit a large number of GP models. In the current literature, many integrated GP models for both quantitative and qualitative factors are proposed [22, 41, 50, 65, 66].

Reference [60] proposed a new method called EzGP to deal with such problems. Let the kth input of the computer emulator be $\mathbf{w}_k = \left(\mathbf{x}_k^T, \mathbf{z}_k^T \right)^T$, where $\mathbf{x}_k = \left(x_{k1}, \ldots, x_{kp} \right)^T$ is the continuous part of input as mentioned in the previous sections and $\mathbf{z}_k = \left(z_{k1}, \ldots, z_{kq} \right)^T \in \mathbb{N}^q$ is the qualitative part of the input, where $k = 1, \ldots, n$. The EzGP method is inspired by the idea of Analysis of Variance (ANOVA) where quantitative and qualitative inputs are jointly modeled as follows:

$$y(\mathbf{w}) = \mu + Z_{\mathbf{z}}(\mathbf{x}), \tag{12}$$

which suggests that for any given level combination of qualitative factors, $y(\mathbf{w})$ is a GP. Specifically, they considered the following additive model structure:

$$Z_{\mathbf{z}}(\mathbf{x}) = Z_0(\mathbf{x}) + Z_{z^{(1)}}(\mathbf{x}) + \cdots + Z_{z^{(q)}}(\mathbf{x}), \tag{13}$$

where Z_0 and $Z_{z^{(h)}}$ for $h = 1, \ldots, q$ are independent GPs with mean zero and some covariance functions. Here, Z_0 plays the role of base GP which takes only quantitative inputs reflecting the intrinsic relation between y and $\mathbf{x}$, and other GPs $Z_{z^{(h)}}$ are the adjustments made to the base GP to reflect the impact of each qualitative factor $z^{(h)}$ for $h = 1, \ldots, q$. The EzGP method can easily deal with heterogeneity in computer models with multiple qualitative factors. Two variants in EzGP are proposed to fit data with high dimensionality or large run sizes, which can achieve high computational efficiency.

3 Designs for Computer Experiments

Computer codes generate outputs in a deterministic manner in computer experiments, meaning the same input returns the same output (no random errors). Latin hypercube designs (LHDs, [38]) are the most popular experimental designs in computer experiments. An n runs and k factors LHD is an $n \times k$ matrix with each column being a random permutation of numbers $1, \ldots, n$. LHDs do not have replicates in each one-dimensional projection. There are various types of optimal LHDs for practical needs, including space-filling LHDs, maximum projection LHDs and orthogonal LHDs.

When we have little or no information about the response surface, it is desirable to have design points as scattered out as possible in the design space for better exploration. Despite LHDs having a uniform one-dimensional projection property, random LHDs may have poor space-filling properties over the entire design space. Figure 4 is an illustrative example with two LHD designs. The LHD in the left panel is concentrated almost entirely on the diagonal, which clearly does not explore the input space sufficiently. The design points in the right panel are scattered out over the entire design space, so this design may provide more reliable information. The maximin distance criterion [25] is a widely used metric for measuring the space-filling property of LHDs. It aims to maximize the minimum distances between design points. Let $\mathbf{X}$ denote an LHD matrix, where the L_q-distance between two runs x_i and x_j of $\mathbf{X}$ is given by $d_q(x_i, x_j) = \left\{ \sum_{k=1}^{m} |x_{ik} - x_{jk}|^q \right\}^{1/q}$, where q is an integer. Two popular choices are $q = 1$ (i.e. the Manhattan distance) and $q = 2$ (i.e. the Euclidean distance). The maximin L_q-distance design has the maximized minimum L_q-distance, i.e. $\max \min d_q(x_i, x_j)$, where $1 \leq i < j \leq n$. Reference [24, 39] further proposed a scalar value to evaluate the maximin distance criterion:

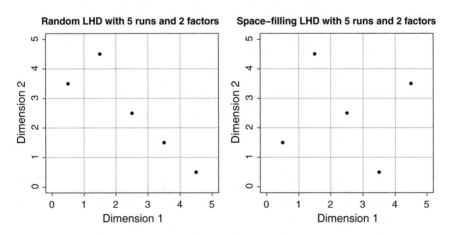

Fig. 4 Latin hypercube designs for size $n = 5$ and $k = 2$

$$\phi_p = \left\{ \sum_{i=1}^{n-1} \sum_{j=i+1}^{n} d_q(x_i, x_j)^{-p} \right\}^{1/p}, \tag{14}$$

where p is a tuning parameter. As $p \to \infty$, the ϕ_p criterion in Eq. (14) is asymptotically equivalent to the Maximin distance criterion, and $p = 15$ is usually sufficient in practice. The LHDs that minimize the ϕ_p criterion are called the maximin distance LHDs.

In the literature, both algebraic constructions [56, 67] and search algorithms [3, 24, 27, 31, 32, 39] are proposed to construct maximin distance LHDs. Algebraic constructions usually require very little computational cost to generate optimal LHDs, which are very attractive for large design sizes. Yet, they are only available for certain design sizes. Search algorithms can generate optimal designs of flexible sizes, but they often require more computation resources to identify optimal LHDs. As there are $(n!)^{k-1}$ possible LHDs with n runs and k factors, search algorithms could become very costly when n and k are large. Here, we will briefly survey some popular construction methods; see [55] for a survey.

Specifically, [56] proposed to generate maximin distance LHDs via good lattice point (GLP) sets [67] and Williams transformation [59]. They proved that the resulting designs of sizes $n \times (n-1)$ (with n being any odd prime) and $n \times n$ (with $2n + 1$ or $n + 1$ being odd prime) are optimal under the maximin L_1-distance criterion. The construction method starts by generating a GLP design, and then use the Williams transformation [59] to improve a linear permuted GLP design. Reference [51] proposed to construct orthogonal array-based LHDs (OALHDs) from existing orthogonal arrays (OAs). The key idea of this construction is to deterministically replace OA entries with a random permutation of LHD elements. OALHDs inherit the properties of OAs and tend to have better space-filling properties compared to random LHDs. Note that the design sizes of OALHDs rely on the existence of corresponding OAs.

Search algorithms should be used to generate optimal LHDs when no construction methods are available. Reference [39] proposed a simulated annealing (SA) algorithm, which randomly exchanges elements to seek improvements over iterations to identify global best LHDs. Following the work of [39] and [51], [31] proposed to construct orthogonal array-based LHDs (OALHDs) using the SA algorithm. They proposed to exchange elements that share the same original OA entry randomly. Reference [27] proposed a multi-objective criterion and developed a modified SA algorithm to generate optimal LHDs having good space-filling properties as well as orthogonality. This algorithm can lead to many good designs, but it is often computationally heavy, since it calculates all average pairwise correlations and row-wise distances at each iteration. Besides these SA-based algorithms, [32] proposed to use a genetic algorithm (GA) for searching optimal designs, which focuses on global best by exchanging random columns between global best and other candidate solutions. In addition, [3] proposed a version of the particle swarm optimization (PSO) algorithm by gradually reducing the Hamming distances between each particle and its personal best (or the global best). Generally speaking, the PSO is recommended for small design sizes ($n \leq 7$) and the GA has better performance for moderate and large design sizes.

Uniform designs (UDs) [11, 13] are another popular type of space-filling designs. There are various measurements of uniformity proposed in the literature, such as the star L_2-discrepancy [57], modified L_2-discrepancy [14] and the centered L_2-discrepancy [15]. The search algorithms mentioned above can be used for identifying UDs.

Maximin distance LHDs have space-filling properties in the full-dimensional space, but their two to $k-1$-dimensional projections may not be space-filling. Reference [26] proposed the maximum projection LHDs (MaxPro LHDs) which enhance the space-filling properties in all possible dimensional projections. Analogous to [39], [26] defined the maximum projection criterion as

$$\min_{\mathbf{X}} \psi(\mathbf{X}) = \left\{ \frac{1}{\binom{n}{2}} \sum_{i=1}^{n-1} \sum_{j=i+1}^{n} \frac{1}{\Pi_{l=1}^{k}(x_{il} - x_{jl})^2} \right\}^{1/k}. \tag{15}$$

LHDs that minimize the ψ values are called MaxPro LHDs. Reference [26] proposed an SA-based search algorithm to identify MaxPro LHDs.

Orthogonal LHDs (OLHDs) are another type of optimal LHDs which aim to minimize the correlations between factors [16, 45, 48]. Two correlation-based criteria are often used to measure designs' orthogonality: the average absolute correlation criterion and the maximum absolute correlation criterion [16], which are defined as

$$\mathrm{ave}(|q|) = \frac{2\sum_{i=1}^{k-1}\sum_{j=i+1}^{k} |q_{ij}|}{k(k-1)} \text{ and } \max|q| = \max_{i,j} |q_{ij}|, \tag{16}$$

where q_{ij} is the correlation between the ith and jth columns in the design matrix. Orthogonal designs may not exist for all sizes. In practice, designs with small $\mathrm{ave}(|q|)$ or $\max|q|$ are preferred.

In the literature, construction methods of OLHDs are widely explored. Specifically, [62] proposed a method to construct OLHDs with run sizes $n = 2^m + 1$ and factor sizes $k = 2m - 2$, where m is any integer no less than 2. Reference [5] extended the work of [62] to accommodate more factors. Reference [45] developed a method based on factorial designs with group rotations for $n = 2^{2^m}$ and $k = 2^m t$, where m is any positive integer and t is the number of rotation groups. Reference [47] improved their earlier work [46] to construct OLHDs with even more flexible run sizes: $n = r2^{c+1}$ or $n = r2^{c+1} + 1$ and $k = 2^c$, where c and r are any two positive integers. Reference [61] proposed to use generalized orthogonal designs to construct OLHDs and nearly orthogonal LHDs (NOLHDs) with $n = 2^{r+1}$ or $n = 2^{r+1} + 1$ and $k = 2^r$, where r is any positive integer. Reference [17] proposed to take advantage of orthogonal matrices and their full fold-overs for constructing OLHDs with $n = 2ak$ runs and k factors, where k is the size of orthogonal matrix and a is any positive integer. Reference [2] implemented the Williams transformation [59] to construct OLHDs with odd prime run-size n and factor-size $k \leq n - 1$. Reference [33] proposed to couple OLHDs or NOLHDs with OAs to accommodate large numbers of factors with fewer runs: n^2 runs and $2fp$ factors, where n and p are design sizes of the OLHDs or NOLHDs and $2f$ is the number of columns in the coupled OA.

4 Discussion

There are many instances in nature where it is either expensive or impossible to conduct a physical experiment. For example, it is prohibitively difficult to conduct a study for investigating the devastation caused by a nuclear explosion. Instances like the formation of a galaxy or the formation of binary black holes cannot be studied through physical experiments. Computer experiments can simulate such phenomena with reasonable accuracy. Although such computer simulators are a lot more desirable than real experiments, they are still computationally expensive. To deal with this problem, scientists use surrogates (emulators) to facilitate the analysis and optimization of complex systems. GPs are widely used as surrogates (or emulators). Space-filling designs, such as LHDs, are often used to reap the benefits of utilizing such surrogates effectively.

Several efficient packages in R are available for fitting the GP model and identifying LHDs. Interested readers can explore different packages for fitting GP: Local Approximate Gaussian Process Regression (laGP) by [19], DiceKriging (Kriging Methods for Computer Experiments) by [42] and GP-fit (Gaussian Processes Modeling) by [36]. For obtaining LHDs with flexible run sizes, packages like Latin Hypercube Designs (LHD) by [54] and Maximin-Distance (Sliced) Latin Hypercube Designs (SLHD) by [44] can be used.

Even though the computing power has increased dramatically over the last few years, handling big data remains a challenging problem. There is an increasing body of literature for computer experiments with large numbers of data points, but the existing literature on large numbers of input variables is still meager. For details, please refer to the review article by [35]. The problem of data reduction is an active area of research among statisticians and computer scientists, and much progress needs to be done in this area. Recent work on this includes techniques like kernel handling [4] and support points [37].

Different Bayesian approaches for analyzing computer experiments have been discussed in the literature, particularly in the context of uncertainty quantification, but most of them are difficult to implement and time-consuming [18, 28]. To solve this problem, we need more advanced techniques. Another topic of active research is to incorporate qualitative input variables. Many practical applications have both quantitative and qualitative inputs, e.g. the data center computer experiment [41] and the study of high-performance computing systems [64]. However, traditional GP modeling is designated for only quantitative inputs, since its covariance function of responses is constructed under the continuous input space with proper distance metrics. More effective techniques and algorithms need to be developed that can accommodate qualitative inputs and one such recent work is [60].

Finally, there is vast existing literature on continuous response, but there are many instances where the response is binary or non-continuous. For example, binary black hole formation [34] or computer experiments with binary time series have non-Gaussian observations [49]. For handling high-dimensional input parameter space, input variables with non-continuous characteristics and non-Gaussian observations, new techniques and algorithms need to be developed.

References

1. Bhattacharjee NV, Ranjan P, Mandal A, Tollner E (2019) A history matching approach for calibrating hydrological models. Environ Ecol Stat 26:87–105
2. Butler NA (2001) Optimal and orthogonal Latin hypercube designs for computer experiments. Biometrika 88(3):847–857
3. Chen R-B, Hsieh D-N, Hung Y, Wang W (2013) Optimizing Latin hypercube designs by particle swarm. Stat Comput 23(5):663–676
4. Chen Y, Welling M, Smola A (2010) Super-samples from kernel herding. In: UAI
5. Cioppa TM, Lucas TW (2007) Efficient nearly orthogonal and space-filling Latin hypercubes. Technometrics 49(1):45–55
6. Cui J, Krems RV (2015) Gaussian process model for collision dynamics of complex molecules. Phys Rev Lett 115(7)
7. Deisenroth M, Fox D, Rasmussen C (2015) Gaussian processes for data-efficient learning in robotics and control. IEEE Trans Pattern Anal Mach Intell 37:408–423
8. Deng F, Deng Q, Shen S (2018) A three-dimensional mixed finite element for flexoelectricity. J Appl Mech 85(3)
9. Doob JL, Doob JL (1953) Stochastic Processes. Wiley, Probability and Statistics Series
10. Ek CH, Torr PH, Lawrence ND (2008) Gaussian process latent variable models for human pose estimation. In: Popescu-Belis A, Renals S, Bourlard H (eds) Machine learning for multimodal interaction. Berlin, Heidelberg, Springer Berlin Heidelberg, pp 132–143
11. Fang K-T (1980) The uniform design: application of number-theoretic methods in experimental design. Acta Math Appl Sin 3(4):9
12. Fang KT, Li R, Sudjianto A (2005) Design and modeling for computer experiments. Chapman and Hall/CRC
13. Fang KT, Li R, Sudjianto A (2005) Design and modeling for computer experiments. CRC Press
14. Fang KT, Lin DK, Winker P, Zhang Y (2000) Uniform design: theory and application. Technometrics 42(3):237–248
15. Fang K-T, Ma C-X, Winker P (2002) Centered L_2-discrepancy of random sampling and Latin hypercube design, and construction of uniform designs. Math Comput 71(237):275–296
16. Georgiou SD (2009) Orthogonal Latin hypercube designs from generalized orthogonal designs. J Stat Plan Inference 139(4):1530–1540
17. Georgiou SD, Efthimiou I (2014) Some classes of orthogonal Latin hypercube designs. Stat Sin 24(1):101–120
18. Gramacy RB, Lee HK (2007) Bayesian treed gaussian process models with an application to computer modeling. J Am Stat Assoc 103:1119–1130
19. Gramacy RB (2019) laGP: large-scale spatial modeling via local approximate gaussian processes in R. J Stat Softw 72(1):1–46
20. Gramacy RB (2020) Surrogates: gaussian process modeling, design and optimization for the applied sciences. Chapman Hall/CRC, Boca Raton, Florida. http://bobby.gramacy.com/surrogates/
21. Gramacy RB, Apley DW (2015) Local gaussian process approximation for large computer experiments. J Comput Graph Stat 24(2):561–578

22. Han G, Santner TJ, Notz WI, Bartel DL (2009) Prediction for computer experiments having quantitative and qualitative input variables. Technometrics 51(3):278–288
23. Hensman J, Fusi N, Lawrence ND (2013) Gaussian processes for big data. In: Technical report UAI-P-2013-PG-282-290, twenty-ninth conference on uncertainty in artificial intelligence (UAI2013)
24. Jin R, Chen W, Sudjianto A (2005) An efficient algorithm for constructing optimal design of computer experiments. J Stat Plan Inference 134(1):268–287
25. Johnson ME, Moore LM, Ylvisaker D (1990) Minimax and maximin distance designs. J Stat Plan Inference 26(2):131–148
26. Joseph VR, Gul E, Ba S (2015) Maximum projection designs for computer experiments. Biometrika 102(2):371–380
27. Joseph VR, Hung Y (2008) Orthogonal-maximin Latin hypercube designs. Stat Sin 171–186
28. Kennedy MC, O'Hagan A (2001) Bayesian calibration of computer models. J R Stat Soc: Series B (Statistical Methodology) 63(3):425–464
29. Kidder LE, Scheel MA, Teukolsky SA, Carlson ED, Cook GB (2000) Black hole evolution by spectral methods. Phys Rev D 62:084032
30. Kruckow M, Tauris T, Langer N, Kramer M, Izzard R (2018) Progenitors of gravitational wave mergers: binary evolution with the stellar grid-based code combine. Mon Not R Astron Soc 481:1908–1949
31. Leary S, Bhaskar A, Keane A (2003) Optimal orthogonal-array-based Latin hypercubes. J Appl Stat 30(5):585–598
32. Liefvendahl M, Stocki R (2006) A study on algorithms for optimization of Latin hypercubes. J Stat Plan Inference 136(9):3231–3247
33. Lin CD, Mukerjee R, Tang B (2009) Construction of orthogonal and nearly orthogonal Latin hypercubes. Biometrika 96(1):243–247
34. Lin L, Bingham D, Broekgaarden F, Mandel I (2021) Uncertainty quantification of a computer model for binary black hole formation
35. Liu H, Ong YS, Shen X, Cai J (2020) When gaussian process meets big data: a review of scalable GPS. IEEE Trans Neural Netw Learn Syst 31:4405–4423
36. MacDonald B, Ranjan P, Chipman H (2015) GPfit: an R package for fitting a gaussian process model to deterministic simulator outputs. J Stat Softw 64(12):1–23
37. Mak S, Joseph VR (2018) Support points. Ann Stat
38. McKay MD, Beckman RJ, Conover WJ (1979) Comparison of three methods for selecting values of input variables in the analysis of output from a computer code. Technometrics 21(2):239–245
39. Morris MD, Mitchell TJ (1995) Exploratory designs for computational experiments. J Stat Plan Inference 43(3):381–402
40. Nocedal J, Wright S (2006) Numerical optimization. Springer series in operations research and financial engineering. Springer, New York
41. Qian PZ, Wu H, Wu CJ (2008) Gaussian process models for computer experiments with qualitative and quantitative factors. Technometrics 50(3):383–396
42. Roustant O, Ginsbourger D, Deville Y (2012) DiceKriging, DiceOptim: two R packages for the analysis of computer experiments by kriging-based metamodeling and optimization. J Stat Softw 51(1):1–55
43. Sacks J, Welch WJ, Mitchell TJ, Wynn HP (1989) Design and analysis of computer experiments. Stat Sci 409–423
44. Brenneman WA, Ba S, Myers WR (2015) Slhd: maximin-distance (sliced) Latin hypercube designs. Technometrics 57(4):479–487
45. Steinberg DM, Lin DK (2006) A construction method for orthogonal Latin hypercube designs. Biometrika 93(2):279–288
46. Sun F, Liu MQ, Lin DK (2009) Construction of orthogonal Latin hypercube designs. Biometrika 96(4):971–974
47. Sun F, Liu MQ, Lin DK (2010) Construction of orthogonal Latin hypercube designs with flexible run sizes. J Stat Plan Inference 140(11):3236–3242

48. Sun F, Tang B (2017) A general rotation method for orthogonal Latin hypercubes. Biometrika 104(2):465–472
49. Sung CL, Hung Y, Rittase W, Zhu C, Jeff Wu CF (2017) A generalized gaussian process model for computer experiments with binary time series. J Am Stat Assoc 115:945–956
50. Swiler LP, Hough PD, Qian P, Xu X, Storlie C, Lee H (2014) Surrogate models for mixed discrete-continuous variables. In: Constraint programming and decision making, Studies in computational intelligence, pp 181–202. Springer
51. Tang B (1993) Orthogonal array-based Latin hypercubes. J Am Stat Assoc 88(424):1392–1397
52. Velásquez RM, Lara JV (2020) Forecast and evaluation of covid-19 spreading in USA with reduced-space gaussian process regression. Chaos Solitons Fractals 136:109924
53. Vernon I, Goldstein M, Bower R (2009) Galaxy formation: a bayesian uncertainty analysis. Bayesian Anal 5:619–669
54. Wang H, Xiao Q, Mandal A (2020) LHD: Latin Hypercube Designs (LHDs) algorithms. R package version 1.2.0
55. Wang H, Xiao Q, Mandal A (2021) Musings about constructions of efficient Latin hypercube designs with flexible run-sizes
56. Wang L, Xiao Q, Hongquan X (2018) Optimal maximin L_1-distance Latin hypercube designs based on good lattice point designs. Ann Stat 46(6B):3741–3766
57. Warnock TT (1972) Computational investigations of low-discrepancy point sets. In: Applications of number theory to numerical analysis, pp 319–343. Elsevier
58. Williams D, Heng IS, Gair J, Clark JA, Khamesra B (2020) Precessing numerical relativity waveform surrogate model for binary black holes: a gaussian process regression approach. Phys Rev D 101:063011. https://journals.aps.org/prd/abstract/10.1103/PhysRevD.101.063011
59. Williams EJ (1949) Experimental designs balanced for the estimation of residual effects of treatments. Aust J Chem 2(2):149–168
60. Xiao Q, Mandal A, Lin CD, Deng X (2021) Ezgp: easy-to-interpret gaussian process models for computer experiments with both quantitative and qualitative factors. SIAM/ASA J Uncertain Quantif 9(2):333–353
61. Yang J, Liu MQ (2012) Construction of orthogonal and nearly orthogonal Latin hypercube designs from orthogonal designs. Stat Sin 433–442
62. Ye KQ (1998) Orthogonal column Latin hypercubes and their application in computer experiments. J Am Stat Assoc 93(444):1430–1439
63. Zhang MM, Dumitrascu B, Williamson SA, Engelhardt BE (2019) Sequential gaussian processes for online learning of nonstationary functions. arXiv:abs/1905.10003
64. Zhang Q, Chien P, Liu Q, Xu L, Hong Y (2020) Mixed-input gaussian process emulators for computer experiments with a large number of categorical levels. J Qual Technol 1–11
65. Zhang Y, Tao S, Chen W, Apley DW (2020) A latent variable approach to gaussian process modeling with qualitative and quantitative factors. Technometrics 62(3):291–302
66. Zhou Q, Qian PZ, Zhou S (2011) A simple approach to emulation for computer models with qualitative and quantitative factors. Technometrics 53(3):266–273
67. Zhou Y, Hongquan X (2015) Space-filling properties of good lattice point sets. Biometrika 102(4):959–966
68. Zhu Y, Fujimura K (2010) A bayesian framework for human body pose tracking from depth image sequences. Sensors (Basel, Switzerland) 10:5280–5293

DNA Molecule Confined in a Cylindrical Shell: Effect of Partial Confinement

Neha Mathur, Arghya Maity, and Navin Singh

Abstract To study the behaviour of DNA molecules during the encapsulation process is a topic of intense research. In the present work, we investigate the stability of the double-stranded DNA molecule of different lengths in a confined shell using a statistical model. The DNA molecules of different lengths are confined in a cylindrical shell either partially or entirely. We consider cylinders of different sizes and study the effect of the size of the cylinder on the microscopic details of the opening of the base pairs.

1 Introduction

Deoxyribonucleic acid (DNA) is one of the interesting and complex biomolecules. It is central to all living beings and contains the information needed for birth, development, living and probably sets the average life. The average length of a DNA molecule ranges from 2 μm (viruses) to $\sim 10^7$ μm for more complex organisms [1]. It is an intense field of research to study the evolution of the molecule and how it acquired the ability to store and transmit genetic information. The recent progress in genome mapping and the availability of experimental techniques to study the physical properties of a single molecule indeed made the field very active from both biological and physical points of view. The genetic information of the entire organism is coded in the sequence of four nucleotides which are named as *Adenine (A)*, *Guanine (G)*, *Cytosine (C)* and *Thymine (T)*. Soon after discovering the molecule's structure, people studied the stability of the DNA molecule in a thermal ensemble. The disruption of double-stranded DNA into single-stranded DNA is known as *DNA melting*

N. Mathur · N. Singh (✉)
BITS Pilani, Pilani campus, Pilani, India
e-mail: navin@pilani.bits-pilani.ac.in

N. Mathur
e-mail: p20180045@pilani.bits-pilani.ac.in

A. Maity
Harish Research Institute, Allahabad, India

© The Author(s), under exclusive license to Springer Nature Singapore Pte Ltd. 2022
R. Srinivas et al. (eds.), *Advances in Computational Modeling and Simulation*,
Lecture Notes in Mechanical Engineering,
https://doi.org/10.1007/978-981-16-7857-8_3

31

or *denaturation*. We can achieve the melting by changing the pH of the solution as well as by pulling either of the strands by a mechanical device [2–4]. The research is going on to use DNA molecules to develop molecular motors, biomedicines and origami formations. In the gene therapy [5] and nanorobotics [6], DNA degradation is a major challenge. It may be due to the chemical breakdown or due to mechanical forces [7]. In gene therapy techniques, DNA is protected by a physical barrier. To protect it from the damage, the DNA can be confined within gel [8, 9], polymeric nanocapsules(micelles) [10] and microparticles. People across the globe are using many elegant and versatile techniques to encapsulate DNA molecules [11].

It has been observed that during the complete process, the processes of encapsulation and release of DNA are inversely related. A better encapsulation process needs a balance between these two sub-processes. Researchers are trying to balance it in numerous ways. In one of the methods, short DNA is encapsulated in a spherical inorganic nanoshell with an overall thickness of ~10 nm [12]. Some researchers found carbon nanotubes as a potential candidate to encapsulate the DNA molecules [13].

In vivo, the DNA molecule is confined in a limited space such as the cell chamber or a channel and is in highly dense solvent conditions [14–17]. The thermal properties of DNA molecules are sensitive to the nature of confined space [18]. DNA packing in eukaryotic chromosomes and viral capsids are some examples in which the activities of the biomolecules depend on the nature of confinement.

In vitro, in most of the studies, the DNA is confined either in a spherical geometry or in a rectangular geometry [14]. Although some significant work is done, our knowledge about the nature of confinement is limited [19–22]. Motivated by the experiments on DNA confined in different geometries, we attempt, here, to understand the thermal stability of a DNA molecule partially confined in a cylindrical shell using a statistical model. It helps to understand the details of confinement and its role in the dynamics of the molecule. The present work is divided into the following sections: in Sect. 2, we introduce the statistical model, and in Sect. 3 we show the results obtained. In the end, we close the manuscript with a brief discussion and future scope of the work (Sect. 4).

2   Model and Methods

We consider the well-known Peyrard-Bishop-Dauxois model (PBD) to investigate the effect of confinement on the thermal denaturation of DNA molecules. The model is quasi-one-dimensional in nature and considers the dynamics of the system through the stretching of the hydrogen bonds between the bases in a pair [23, 24]. Although it ignores the effect of the molecule's helicoidal nature and the solvent effect of the solution, the model has enough details to explain the denaturation/unzipping process of DNA molecules. The interactions in the DNA, containing N base pairs, are represented as

$$H = \sum_{i=1}^{N} \left[\frac{p_i^2}{2m} + V_M(y_i) \right] + \sum_{i=1}^{N-1} \left[W_S(y_i, y_{i+1}) \right], \tag{1}$$

here y_i represents the separation between two bases in a pair. The separation $y_i = 0.0$ Å refers to the equilibrium position of two bases in a pair. The first term of the model is the momentum term which is $p_i = m\dot{y}_i$. We have taken same reduced mass $m = 300$ amu for both the AT and GC base pairs [23]. The stacking interaction between the nearest base pairs along the chain is represented as

$$W_S(y_i, y_{i+1}) = \frac{k}{2}(y_i - y_{i+1})^2[1 + \rho e^{-b(y_i + y_{i+1})}]. \tag{2}$$

Here, k represents the single-strand elasticity. The term ρ represents the anharmonicity in the strand elasticity while the parameter, b, describes its range. In the present work, we choose model parameters $k = 0.02$ eVÅ^{-2}, $\rho = 5.0$ and $b = 0.35$ Å^{-1}. In our earlier works, we have shown that values of k and ρ define the sharpness in the transition from double strand to single strand [25, 26]. The pairing between the bases in a pair is represented by the Morse potential as

$$V_M(y_i) = D_i(e^{-a_i y_i} - 1)^2. \tag{3}$$

Here D_i represents the potential depth, and a_i represents the inverse of the width of the potential well. These two parameters have a crucial role in DNA denaturation. From previous results, we know that the bond strengths of these two pairs are in an approximate ratio of 1.25–1.5 as the GC pairs have three while AT pairs have two hydrogen bonds [27, 28]. The potential parameters are taken as $a_{AT} = 4.2$ Å^{-1}, $a_{GC} = 1.5 \times a_{AT}$ and $D_{AT} = 0.05$ eV while $D_{GC} = 1.5 \times D_{AT}$. The complete set of parameters are $D_{AT} = 0.05$ eV, $D_{GC} = 0.075$ eV, $a_{AT} = 4.2$ Å^{-1}, $a_{GC} = 6.3$ Å^{-1}, $\rho = 5.0$, $\kappa = 0.02$ eV/Å^2 and $b = 0.35$ Å^{-1}. The model parameters are tuned in such a way that the melting temperature of 12 base pairs chain is between 300–350 K. We can study the thermodynamics of the transition by evaluating the partition function. For a sequence of N base pairs, the canonical partition function can be written as

$$Z = \int \prod_{i=1}^{N} \{dy_i dp_i \exp(-\beta H)\} = Z_p Z_c, \tag{4}$$

where Z_p corresponds to the momentum part of the partition function and is equal to $(2\pi m k_B T)^{N/2}$. The configurational part of the partition function, Z_c, is defined as

$$Z_c = \int_{-\infty}^{\infty} dy_1 e^{-\frac{1}{2}\beta V(y_1)} \left[\prod_{i=1}^{N-1} dy_i K(y_i, y_{i+1}) \right] e^{-\frac{1}{2}\beta V(y_N)} dy_N, \tag{5}$$

where $K(y_i, y_{i+1}) = e^{-\frac{\beta}{2}[V(y_i)+V(y_{i+1})+2W(y_i, y_{i+1})]}$.

If one considers the DNA of a homogeneous sequence and periodic boundary condition, one can evaluate the configurational partition function, Z_c, using the transfer integral (TI) technique. For a chain with a random sequence of AT and GC pairs and open boundaries, partition function calculation is a little bit tricky. In the past, various researchers have addressed solving the chain's partition function with heterogeneous sequence and open boundaries. Also, the partition function in the PBD model is divergent in nature. To overcome this problem, an upper cut-off for the integration needs to be set up [29–31]. In our previous studies, we found that an upper cut-off of 200 Å is sufficient to overcome the divergence issue of the partition function. The lower limit of integration is set as –5.0 Å [26, 29]. Once we find the proper cut-offs, the task is to discretize the integral in Eq. 5. In order to get a precise value of melting temperature (T_m), we have observed that Gaussian quadrature is the most effective quadrature. We have found that discretization of the space with 900 points is sufficient to get an accurate value of T_m. Once we can evaluate the partition function, we can determine the thermodynamic quantities of interest by evaluating the Helmholtz free energy of the system. We define the Helmholtz free energy per base pair as

$$f(T) = -\frac{1}{2}k_BT \ln(2\pi mk_BT) - \frac{k_BT}{N} \ln Z_c. \tag{6}$$

The specific heat, C_v, of the system, in the thermal ensemble, is evaluated by taking the second derivative of the free energy as $C_v = -T(\partial^2 f/\partial T^2)$. We calculate the melting temperature (T_m) of the chain from the peak in the specific heat curve. Other quantity of interest is the average separation, $\langle y_j \rangle$, of the jth pair of the chain, which is given by

$$\langle y_j \rangle = \frac{1}{Z} \int \prod_{i=1}^{N} y_j \exp(-\beta H)dy_i. \tag{7}$$

This work studies the stability of DNA molecules confined in a cylindrical shell of different lengths. We chose DNA of different lengths and confined them in a cylinder of different dimensions (as shown in Fig. 1). How to realise the confinement in the model? In the calculation of partition function, we restrict the configuration space of the system as shown in Fig. 1. Since the confinement restricts the stretching of two

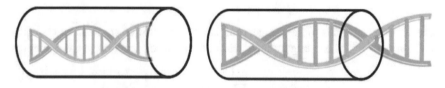

Fig. 1 The schematic representation of the DNA molecule confined in a cylindrical shell. The r is the distance of the confined wall from the DNA strand. The radius of the cylinder is $R_c = r + $ DNA radius (10 Å). The DNA is confined in the shell either completely or partially

bases in the pair, we represent it through the upper cut-off of the integral in the model. For a pair that is unconfined, the upper limit of the integration is 200 Å, while for the pair that is confined, the upper limit of integral is r Å. For the chain confined in the cylinder of radius 10 Å, the upper limit of integration is 10 Å. Since the confinement affects outward stretching, we assume that there is no change in the lower limit of integral [20, 21]. We have assumed that the DNA is not interacting with the walls of the cylinder. We calculate the partition function with the modified configuration space and evaluate all the system's thermodynamical properties for both geometries.

3   Results

We consider the first 12 base pairs of the phage-λ DNA chain and repeat the sequence to form other sequences for our studies. The sequence of the chain is $-GGGGAAAAGGGG-$. Each chain is confined in the cylinder of lengths 20, 50 and 150 bp. Here we consider the length of the cylinder in terms of base pairs for mathematical simplicity. We calculate free energy per base pair by evaluating the partition function and hence the specific heat as a function of temperature. Through the peak in the specific heat, we identify the melting temperature T_m of the chain at different diameters of the cylinder as shown in Fig. 2. Once we find the melting temperature of the system, we change the length of the DNA molecule. In our previous works, we have shown the effect of confinement on the melting temperature of DNA [20, 21]. In the current work, our interest is to investigate the change in the melting temperature of the partially confined DNA molecule. The partially confined system is very close to the translocation of DNA through the cell. First we consider the cylinder of fixed length $l = 20$ and radial distance $r = 10$ Å. Please note that $r = 10$ Å means the cylinder is of radius $R_c = r +$ DNA radius. We calculate the change in the melting temperature of the system with the increasing radial distance.

Fig. 2   The change in specific heat with the temperature of the system. The plot is for DNA molecules of lengths 12, 36 & 72 base pairs confined in a cylinder of length 50 base pairs and radius of 10 Å

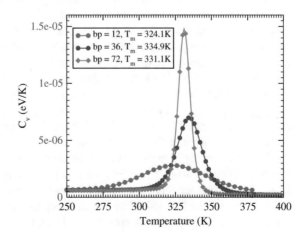

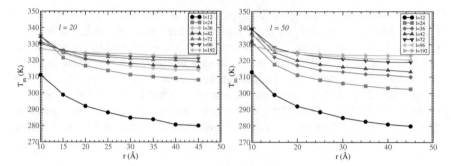

Fig. 3 The melting temperature of the DNA molecules that is confined in a cylinder of different lengths. We consider lengths as 20 and 50 base pairs. The plots show the changes in the T_m with the increasing radius of the cylinder for the chains 12–192 base pairs

Fig. 4 The melting temperature of the DNA molecules that is confined in a cylinder of different lengths. We consider lengths as 150 base pairs. The plots show the changes in the T_m with the increasing radius of the cylinder for the chains 12–192 base pairs

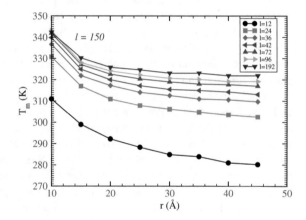

The results are shown in Fig. 3. From the plots, it is clear that the melting temperature of the DNA molecule decreases with increasing radial distance, r. The reason for the decrease is the available space to the molecule. The stability seems to be as per the expectation; however, there is a striking and interesting difference at r=10 Å for the molecules of length larger than 20 base pairs. We find that T_m is almost the same for 36 and 42 base pairs chain while it lowers for the chains of 72, 96 and 192 base pairs. We observe a similar kind of feature for the cylinder of different lengths (see Fig. 3). However, for the molecules confined in a cylinder of length 150 base pairs (see Fig. 4), the previously observed pattern is lost. The reason for the change is the confinement length. The entropy, in this case, is suppressed due to the large size of the cylinder. We plot the change in the value of T_m with different chain lengths for three different cylinder lengths to understand the process. By varying the size of DNA molecules from 12 to 192, we calculate the melting temperature of each chain. The results are shown in Fig. 5. As the chain length increases the melting temperature, the system's stability increases up to a certain length and decreases. When the size of the molecule crosses this limit, two sections of the chain behave

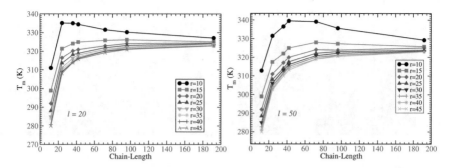

Fig. 5 The melting temperature of the DNA molecules that is confined in a cylinder of different lengths. We consider lengths as 20 and 50 base pairs. The plots show the changes in the T_m with the increasing chain length for different radii of the cylinder

Fig. 6 The melting temperature of the DNA molecules that is confined in a cylinder of different lengths. We consider lengths as 150 base pairs. The plots show the changes in the T_m with the increasing chain length for different radii of the cylinder

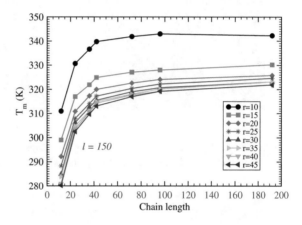

differently. A section that is free to move hence more entropic, and there is a section of the chain that is still confined. The unconfined or free section of the molecule has sufficient entropy to drive the system from a zipped state to unzipped state. From Fig. 6, we find an increase in the melting temperature for $r = 10$ Å. The point to note is that the increase is up to the length of the cylinder, while after that, the T_m decreases. However, the pattern is not universal. For a larger cylinder radius, the T_m increases and then saturates to a value. For a cylinder of length $l = 150$ bp, even for the $r = 10$ Å, there is no point of inflexion. The model we consider here assumes DNA in ladder form. For our studies, we ignore the change in the persistence length of confined DNA [32].

To get a deeper understanding, we calculate the average separation of DNA molecules of length 12, 36 and 72 base pairs. Each chain is confined in two kinds of cylinders. One of length 20 bp and radius 10 Å, while the other is of the same length but different radial distance $r = 40$ Å. The average separation, $\langle y_i \rangle$, of the chain is defined using Eq. 7. The results are shown in Figs. 7, 8 and 9. Some exciting features of the opening of the DNA molecules are visible in these plots. The opening is more

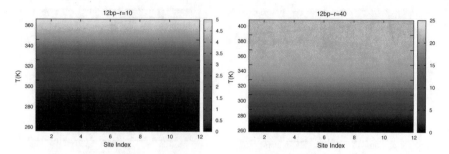

Fig. 7 The density plots showing the change in the average separation, $\langle y_j \rangle$ for DNA of 12 base pairs confined in a cylinder of length 20 base pairs and radii $r = 10$ and $r = 40$ Å

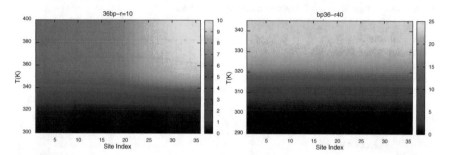

Fig. 8 The density plots showing the change in the average separation, $\langle y_j \rangle$, for DNA of 36 base pairs confined in a cylinder of length 20 base pairs and radial distance $r = 10$ and $r = 40$ Å

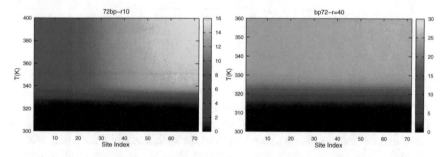

Fig. 9 The density plots showing the change in the average separation, $\langle y_j \rangle$, for DNA of 72 base pairs confined in a cylinder of length 20 base pairs and radial distance $r=10$ and $r = 40$ Å

or less homogeneous for the completely confined chains. For the chain of 36 base pairs and $r = 10$ Å, the chain opens from the free end (see the yellow colour in Fig. 8). For $r = 40$ Å, the opening is little homogeneous. In this condition, the molecule has sufficient space to move now; hence, the effect of confinement is weak in this case. We observe a very similar kind of pattern in the chain of 72 base pairs. When the chain is confined in a radial distance of 10 Å, up to 20 base pairs, the maximum length up to which the base pairs can move is 10 Å, while for a larger radius, this heterogeneity disappears.

4 Conclusions

We have studied the stability of phage-λ DNA molecule partially confined in a cylindrical shell using the PBD model. The DNA molecule is partially inside the cell during the translocation process. Thus, different segments of the same DNA molecule experience different forces due to the neighbouring molecules. In the current manuscript, our focus was to understand the stability of DNA molecule, that is partially confined in a cylinder. We have considered DNA of different lengths and confined them in a cylinder of different radii and lengths. With the variation in the radius and length of the cylinder, we attempt to understand the opening of a heterogeneous DNA molecule confined in the cylinder. By calculating $\langle y_j \rangle$, we have found that the DNA confined in a cylinder experiences a uniform suppression in the system's entropy; however, the opening of partially confined DNA is more interesting. We want to study the effect of a force applied on the DNA molecule confined in a cylindrical shell in future work.

Acknowledgements We acknowledge the financial support from the Department of Science and Technology (EMR/002451), New Delhi.

References

1. Watson JD, Crick F (1953) Nature 171:737
2. Sadhukhan P, Bhattacharjee SM (2014) Indian J Phys 88(9):895. https://doi.org/10.1007/s12648-014-0489-3
3. Hatch K, Danilowicz C, Coljee V, Prentiss M (2008) Nucl Acids Res 36(1):294. https://doi.org/10.1093/nar/gkm1014
4. Kumar S, Li MS (2010) Phys Rep 486(1–2):1. https://doi.org/10.1016/j.physrep.2009.11.001
5. Putnam DJ (2006) Nature materials 5:1476. https://doi.org/10.1038/nmat1645
6. Yurke B, Turberfield AJ, Jr APM, Simmel FC, Neumann JL (2000) Nature 406:605. https://doi.org/10.1038/35020524
7. Murphy JC, Cano T, Fox GE, Willson RC (2006) Biotechnology progress 22(2):519. https://doi.org/10.1021/bp050422i
8. Goh SL, Murthy N, Xu M, Fréchet JMJ (2004) Bioconjugate Chem 15(3):467. https://doi.org/10.1021/bc034159n

9. des Rieux A, Shikanov A, Shea LD (2009) J Control Release 136(2):148. https://doi.org/10.1016/j.jconrel.2009.02.004

10. Csaba N, Caamaño P, Sánchez A, Domínguez F, Alonso MJ (2005) Biomacromolecules 6(1):271. https://doi.org/10.1021/bm049577p

11. Dimitrov IV, Petrova EB, Kozarova RG, Apostolova MD, Tsvetanov CB (2011) Soft Matter 7:8002. https://doi.org/10.1039/C1SM05805C

12. Park DH, Kim JE, Oh JM, Shul YG, Choy JH (2010) J Am Chem Soc 132(47):16735. https://doi.org/10.1021/ja105809e PMID: 20845909

13. Cruz FJAL, de Pablo JJ, Mota JPB (2014) J Chem Phys 140(22):225103. https://doi.org/10.1063/1.4881422

14. Li H, Wang Z, Li N, He X, Liang H (2014) J Chem Phys 141(4):044911. https://doi.org/10.1063/1.4891219

15. Kumar S, Kumar S, Giri D, Nath S (2017) Eur Lett 118(2):28001

16. Turner SWP, Cabodi M, Craighead HG (2002) Phys Rev Lett 88:128103. https://doi.org/10.1103/PhysRevLett.88.128103

17. Singh A, Singh N (2017) Phys Chem Chem Phys 19:19452. https://doi.org/10.1039/C7CP03624H

18. Akabayov B, Akabayov S, Lee S, Wagner G, Richardson C (2013) Nat Commun 4

19. Wanunu M (2012) Phys Life Rev 9(2):125

20. Maity A, Singh A, Singh N (2019) EPL (Eur Lett) 127(2):28001. https://doi.org/10.1209/0295-5075/127/28001

21. Maity A, Singh N (2020) Eur Biophys J 49:561. https://doi.org/10.1007/s00249-020-01462-9

22. Zoli M (2020) EPL (Eur Lett) 130(2):28002. https://doi.org/10.1209/0295-5075/130/28002

23. Peyrard M, Bishop AR (1989) Phys Rev Lett 62(23):2755. https://doi.org/10.1103/PhysRevLett.62.2755

24. Dauxois T, Peyrard M, Bishop AR (1993) Phys Rev E 47(1):R44. https://doi.org/10.1103/PhysRevE.47.R44

25. Cocco S, Monasson R (1999) Phys Rev Lett 83(24):5178. https://doi.org/10.1103/PhysRevLett.83.5178

26. Singh N, Singh Y (2005) Eur Phys J E 17(1):7. https://doi.org/10.1140/epje/i2004-10100-7

27. Zoli M (2021) J Chem Phys 154(19):194102. https://doi.org/10.1063/5.0046891

28. Singh N, Singh Y (2001) Phys Rev E 64(4):042901. https://doi.org/10.1103/PhysRevE.64.042901

29. Zhang Yl, Zheng WM, Liu JX, Chen YZ (1997) Phys Rev E 56(6):7100. https://doi.org/10.1103/PhysRevE.56.7100

30. van Erp TS, Cuesta-Lopez S, Peyrard M (2006) Eur Phys J E 20(4):421. https://doi.org/10.1140/epje/i2006-10032-2

31. Singh A, Singh N (2015) Phys Rev E 92:032703. https://doi.org/10.1103/PhysRevE.92.032703

32. Cifra P, Benkova Z, Bleha T (2010) Phys Chem Chem Phys 12:8934. https://doi.org/10.1039/B923598A

Efficient Physics Informed Neural Networks Coupled with Domain Decomposition Methods for Solving Coupled Multi-physics Problems

Long Nguyen, Maziar Raissi, and Padmanabhan Seshaiyer

Abstract In this work, we introduce a novel coupled methodology called PINNs-DDM that combines a physics informed neural networks (PINNs) approach with a domain decomposition method (DDM) approach to solve multi-physics problems. The coupled methodology is applied to a variety of benchmark problems and validated against their exact solutions. Motivated by the need to solve coupled problems in enclosed spaces, we consider an application of coupling scalar transport equations to fluid dynamics equations using PINNs-DDM. While the examples and benchmark problems used in this work are in lower dimensions, they provide the necessary insight into the efficiency of the coupled method. It was noted that one of the key applications of the method is its performance for problems with limited training data. The computational results suggest that the method is very robust and can be applied to study complex real-world applications.

1 Introduction

Research in computational mathematics, which comprises modeling, analysis, simulation, and computing has become the foundation for solving most multidisciplinary problems in science and engineer. These real-world problems often involve complex dynamic interactions of multiple physical processes which presents a significant challenge, both in representing the physics involved and in handling the resulting coupled behavior. If the desire to predict and learn from the system is added to the picture,

L. Nguyen · P. Seshaiyer (✉)
George Mason University, Mathematical Sciences, Fairfax, USA
e-mail: pseshaiy@gmu.edu

L. Nguyen
e-mail: lnguye33@gmu.edu

M. Raissi
University of Colorado, Applied Mathematics, Boulder, USA
e-mail: Maziar.Raissi@colorado.edu

© The Author(s), under exclusive license to Springer Nature Singapore Pte Ltd. 2022　　41
R. Srinivas et al. (eds.), *Advances in Computational Modeling and Simulation*,
Lecture Notes in Mechanical Engineering,
https://doi.org/10.1007/978-981-16-7857-8_4

then the complexity increases even further. Hence, to capture the complete nature of the solution to the problem, a coupled multidisciplinary approach is essential.

The efficient solution to a complex coupled system that consists of functionally distinct components is still a challenging problem in computational sciences research. Direct numerical solution of the highly non-linear equations governing even the most simplified models are often challenging. The past few decades have seen significant advances in algorithms for efficient solutions including finite element methods for solving coupled multi-physics problems [1, 3, 4, 18, 19], domain decomposition method (DDM) [10, 11, 29] and multi-fidelity methods and sampling methods for Partial Differential Equations [21, 22]. Recent advances in machine learning (ML) methods, automatic differentiation (AD) [7], and ML libraries such as TensorFlow [20] and Pytorch [17] have made them potentially powerful tools for parameter estimation and data assimilation in multi-physics problems. Recently, Physics-informed (deep) neural networks (PINNs) were used to learn solutions and parameters in partial and ordinary differential equations [24]. These methodologies have helped to make tremendous progress in the development, testing, analysis, implementation, and applications of computational mathematics for simulation, optimization, and control.

Our work in this chapter is motivated by the following multi-physics application. Over the last two years, we have been faced with an unprecedented sequence of events due to COVID-19 that has impacted not only health but also economy, jobs, education, and many other sectors. As emergency efforts resume and continue in the coming weeks, thousands of personnel would need to be transported in passenger and cargo compartments. Many of these passengers would be infected or exposed to the virus and several agencies are already preparing to develop rapid solutions to study the speed of the contagion by understanding the airflow inside an aircraft that is transporting people. One of the principal dynamics involved in this process is the interplay between the flow from the air vents inside the cabins modeled via computational fluid dynamics and the scalar transport that models the concentration of the pathogen. This paper will attempt to create a simplistic model to simulate and predict the mechanisms of this coupled dynamics through a spatial and temporal distribution of airborne infection risk in an enclosed space. One of the new contributions in this work is to re-introduce PINNs and DDM methods and then develop a framework to solve multi-physics problems using a coupled PINNs-DDM methodology.

Our outline of the chapter is as follows. In Sect. 2, we will introduce the models, describe the methods and background. Specifically in Sect. 2.1, we recall the Schwarz Domain Decomposition method and apply it to the Poisson equation. Section 2.2 introduces the physics informed neural networks (PINNs) approach to solving PDEs and is applied to Burger's equation. In Sect. 2.3, we develop a new algorithm to couple PINNs with DDM to create a PINNs-DDM algorithm for multi-physics problems and apply it to various benchmark equations. Finally, in Sect. 3 we present discussion and future work.

2 Models, Methods and Background

In this section, we describe a multi-physics model that couples fluid dynamics that models the air velocity with scalar transport that models the concentration of a particle in an enclosed space such as an aircraft cabin. For simplicity, we will keep the exposition in the paper to one dimension that will help provide an insight into higher dimensions. After a brief introduction of these equations, we introduce DDM through a parallel Schwarz algorithm and apply it to the Poisson equation. Then, we will introduce PINNs and apply it as a forward solver to solve the Burgers equation.

The particle transport that models the concentration of a pathogen [12] that is guided by the air velocity may be modeled using the equation:

$$\frac{\partial \phi}{\partial t} + \frac{\partial}{\partial x_i}(\rho \phi U_i) = \frac{\partial}{\partial x_i}\left(\Gamma_\phi \frac{\partial \phi}{\partial x_i}\right) + S_\phi \tag{1}$$

where ϕ is contaminant concentration (for example, droplets from infected or exposed COVID-19 individuals), t is time, x_i is coordinate, ρ is air density, U_i is air velocity, Γ_ϕ is the diffusion coefficient, and S_ϕ is the mass flow rate of source per unit volume. In this work, we will consider a one-dimensional problem with the following equation for describing the

$$\frac{\partial V}{\partial t} + U \frac{\partial V}{\partial x} = \alpha_V(x)\frac{\partial^2 V}{\partial x^2} + S_V \tag{2}$$

where α_V and S_V are the respective diffusion coefficient and external source term. Applications of fluid dynamics in studying airflow in ventilated enclosed spaces have been studied over last several decades [5, 14, 15]. While turbulence models building on three-dimensional Navier-Stokes equations are typically used to evaluate and design various air distributions, we will use the following one-dimensional viscous Burger's equation in this work for simplicity. This is given by

$$\frac{\partial U}{\partial t} + U \frac{\partial U}{\partial x} = \alpha_U(x)\frac{\partial^2 U}{\partial x^2} + S_U \tag{3}$$

Here, α_U and S_U correspond to the viscocity of the flow and external source term, respectively. Often these partial differential equations (2) and (3) are discretized using appropriate numerical methods (such as, e.g., finite difference, finite elements, finite volumes) that involves solving a non-linear system of equations. Next, we describe a framework that can be used with any of these discretization methods to make the algebraic solution more efficient on parallel computer platforms.

2.1 Domain Decomposition Method (DDM)

Domain decomposition methods were introduced in the nineteenth century by German analyst Herman Schwarz as a way to reformulate and solve any given boundary-value problem on a computational domain that is partitioned into multiple subdomains [27]. This convenient framework and several variations over the past decades has allowed for efficient techniques for solving multi-physics problems that are governed by differential equations of various types in different subregions of the computational domain [2, 6, 8, 9, 26, 30].

To illustrate, DDM, let us consider the Poisson equation $-\Delta u = f$ defined on the computational domain Ω with $u = g$ on the boundary $\partial \Omega$. Let Ω be the union of a disk (Ω_1) and a rectangle (Ω_2) as shown in Fig. 1. The key idea behind the classical Schwarz algorithm is to iteratively solve alternating sub-problems in the domains Ω_1 and Ω_2 until the algorithm converges as follows:

$$-\Delta u_i^{n+1} = f \quad in \quad \Omega_i$$
$$u_i^{n+1} = g \quad on \quad \partial \Omega_i \cap \partial \Omega \setminus \partial \Omega_i \cap \overline{\Omega}_{3-i}$$
$$u_i^{n+1} = u_{3-i}^n \quad on \quad \partial \Omega_i \cap \overline{\Omega}_{3-i}$$

for $i = 1, 2$. Schwarz proved the convergence of the algorithm and thus the well-posedness of the Poisson problem in complex geometries. If the algorithm converges, the solutions $u_1^\infty = u_2^\infty$ in the intersection of the subdomains.

We can extend the Schwarz method to a general differential operator $\mathcal{L}(u) = f$ in Ω with boundary condition $\mathcal{B}(u) = g$ on $\partial \Omega$ as follows:

Algorithm 1: A parallel Domain Decomposition Method for Two Sub-domains Ω_1, Ω_2

Result: Give initial guess u_1^0 on $\partial \Omega_1 \cap \overline{\Omega}_2$ and u_2^0 on $\partial \Omega_2 \cap \overline{\Omega}_1$
initialization;
for $n = 1$ to maxIter **do**

> Solve for u_i^n $(i = 1, 2)$:
> $$\mathcal{L}(u_i^n) = f \quad in \quad \Omega_i$$
> $$\mathcal{B}(u_i^n) = g \quad on \quad \partial \Omega_i \cap \partial \Omega \setminus \partial \Omega_i \cap \overline{\Omega}_{3-i}$$
> $$u_i^n = u_{3-i}^{n-1} \quad on \quad \partial \Omega_i \cap \overline{\Omega}_{3-i}$$
>
> **if** $\|u_i^n - u_{3-i}^n\| \leq Tol$ **then**
> > | STOP;
> **end**

end

To demonstrate the performance of the Schwarz DDM, we consider the Poisson equation in 1-dimension $-\dfrac{d^2 u}{dx^2} = 2$ on the domain (α, β) with $u = -2$ at $\alpha = -1$ and $\beta = 1$. The exact solution is $u(x) = -x^2 - 1$ which is plotted as the bold line in Fig. 2 which shows the convergence of solutions in the two subdomains Ω_1, Ω_2.

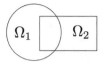

Fig. 1 $\Omega = \Omega_1 \cup \Omega_2$

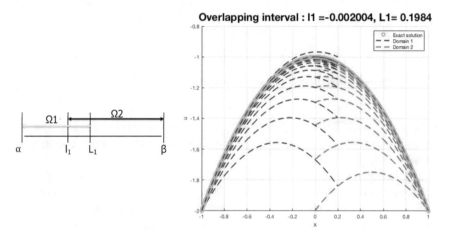

Fig. 2 Computational domain decomposed into two overlapping subdomains (left) and the Schwarz convergence of the computational solution (right)

Next we introduce a framework around probabilistic machine learning to discover governing equations expressed by parametric linear operators.

2.2 *Physics Informed Neural Network (PINNs)*

Physics informed neural networks (PINNs) are deep learning-based techniques [22–24] for solving equations describing multi-physics including ordinary and partial differential, integro-differential and fractional order operators. One of the tools that makes these deep learning methods successful is the use of neural networks which is a system of decisions modeled after the human brain [16].

Consider the illustration shown in Fig. 3. The first layer of perceptrons first weighs and biases the input x. The next layer then will make more complex decisions based on those inputs, until the final decision layer is reached which generates the outputs u. The left part of the figure visualizes a standard neural network parameterized by θ. The middle part in the figure applies the given physical laws to the network. $\mathcal{L}$ and $\mathcal{B}$ are the differential and the boundary operators, respectively. The ODE/PDE data (f, g) are obtained from random sample points. The loss function is computed by evaluating $\mathcal{L}[u]$ and $\mathcal{B}[u]$ on the sample points, which can be done efficiently

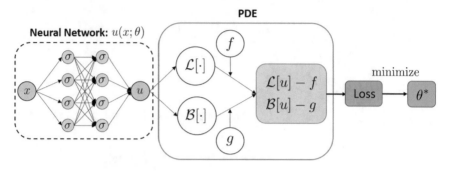

Fig. 3 Illustration of the physics informed neural network approach [24, 28]

through automatic differentiation. Minimizing the loss with respect to the network's parameters θ produces a PINNs $u(x; \theta^*)$, which serves as an approximation to the solution to the ODE/PDE.

In this paper, we implement a physics informed neural network-based approach (PINNs) which makes decisions based on appropriate activation functions depending on the computed bias and weights. The network then seeks to minimize the mean squared error of the regression with respect to the weights and biases by utilizing gradient descent type methods used in conjunction with software such as tensorflow. To demonstrate the performance of the PINNs method, we consider Burger's equation given by for $x \in [0, \pi]$ and $t \in [0, 10]$,

$$u_t = \frac{1}{5}u_{xx} - uu_x + e^{-\frac{2t}{5}}\sin x \cos x \qquad u(t, 0) = u(t, \pi) = 0 \qquad u(0, x) = \sin x$$

In order to create data for our simulation, we use exact solution given by $u(t, x) = e^{-\frac{t}{5}}\sin x$. For our PINNs implementation, we consider 3 hidden layers with size 50 for the neural network. We first choose 200 random data points N_u to estimate u by neural network denoted as u_{pred}. We also choose 200 random data points N_f to estimate PINNs residual. Defining this by,

$$f(x, t) = u_t - \frac{1}{5}u_x x + uu_x - e^{-\frac{2t}{5}}\sin x \cos x$$

we create a Loss function defined by

$$\text{Loss} = \frac{1}{N_u}\sum_{i=1}^{N_u}(u_{train} - u_{pred})^2 + \frac{1}{N_f}\sum_{i=1}^{N_f}(f(x, t))^2 \qquad (4)$$

We train the model via 30,000 iterations using Adam [13] to minimize the Loss given by Eq. (4). Figure 4 shows that the solutions from PINNs taken at various times match

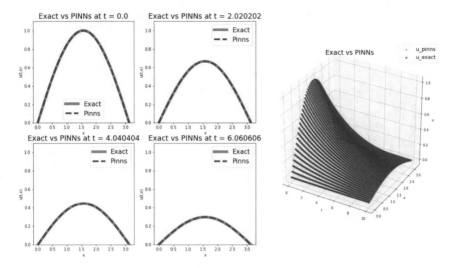

Fig. 4 Comparison of PINNs versus exact solution at different times (left) and for all times and space (right)

very well with the exact solution. The actual relative error in this case is 2×10^{-4}, and the value of the Loss function is 5×10^{-7}.

2.3 A Novel PINNS-DDM Approach

In this section, we will discuss how one can efficiently combine PINNs with DDM which is the novelty of this work, to improve accuracy in the solution methodology.

As described, PINNs is an effective method to solve multi-physics real-world applications modeled via ODE/PDE especially with good amount of training data. However in real-world applications, we often get limited data and DDM combined with PINNs can be an efficient way to solve such problems. The algorithm that will be described next was applied to the benchmark Poisson equation in 1-dimension $-u_{xx} = 2$ on the domain $(-1, 1)$ with $u(-1) = u(1) = -2$ (See Fig. 5).

The result illustrated in Fig. 6 was generated using the following PINNs-DDM algorithm. Specifically, both Neural Networks were chosen to have 3 hidden layers with size of 50 nodes. Choosing random training data points $N_{u_i} = N_{f_i} = 50$ for $i = 1, 2$ as well as $N_{12} = 50$ input points in overlap for training. Running the algorithm

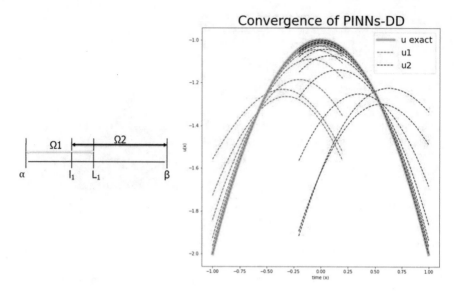

Fig. 5 Performance of PINNs-DDM algorithm

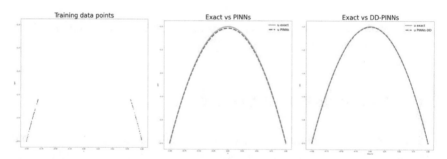

Fig. 6 Training data limited to each sub-domain (left); Approximation using PINNs (middle); Approximation using PINNs-DDM (right)

for maximum number of iterations of 30, 000, we obtained an error in sub-domain one to be 8.1×10^{-5} and in sub-domain two to be 3.8×10^{-5}. Note that the PINNs-DDM is able to reproduce the same solution as in Fig. 2.

Algorithm 2: A novel PINNS-DDM Method for Sub-domains $\Omega_i, i = 1, 2$

Result: Give data and the PDE $\mathcal{L}(u) = f$ with boundary conditions $\mathcal{B}(u) = g$

1 Initialize to create Neural Networks $\mathcal{N}_i$ in $\Omega_i \setminus \Omega_1 \cap \Omega_2$

2 Input the training Data $u_{i,t}^j$, $j = 1..N_{u_i}$

3 Predict the solution $u_{i,p}^j$, $j = 1..N_{u_i}$ using the Neural Network $\mathcal{N}_i$ on $\Omega_i \setminus \Omega_1 \cap \Omega_2$

4 Define the residuals $f_i^j = f_i - \mathcal{L}(u_i^j)$ for $j = 1..N_{f_i}$

5 **while** $n <= $ *maxiter* **do**

  – Define the loss functions Loss_1 and Loss_2 as follows:

$$\text{Loss}_1 = \frac{1}{N_{u_1}} \sum_{j=1}^{N_{u_1}} \left(u_{1,t}^{j,n} - u_{1,p}^{j,n} \right)^2 + \frac{1}{N_{f_1}} \sum_{j=1}^{N_{f_1}} (f_1^{j,n})^2$$

$$\text{Loss}_2 = \frac{1}{N_{u_2}} \sum_{j=1}^{N_{u_2}} \left(u_{2,t}^{j,n} - u_{2,p}^{j,n} \right)^2 + \frac{1}{N_{f_2}} \sum_{j=1}^{N_{f_2}} (f_2^{j,n})^2$$

  – Define the loss function Loss_{12} corresponding to the overlap in the domain decomposion using predicted functions $u_{12,i,p}^{j,n}$ for $j = 1..N_{u_{12}}$

$$\text{Loss}_{12} = \frac{1}{N_{u_{12}}} \sum_{j=1}^{N_{u_{12}}} \left(u_{12,1,p}^{j,n} - u_{12,2,p}^{j,n-1} \right)^2 + \frac{1}{N_{u_{12}}} \sum_{j=1}^{N_{u_{12}}} \left(u_{12,1,p}^{j,n-1} - u_{12,2,p}^{j,n} \right)^2$$

$$+ \frac{1}{N_{f_{12}}} \sum_{j=1}^{N_{f_{12}}} [(f_{12,1}^{j,n})^2 + (f_{12,2}^{j,n})^2]$$

  – Compute total loss $\text{Loss} = \text{Loss}_1 + \text{Loss}_2 + \text{Loss}_{12}$
  – Train the Loss function using Adam's method [13]
  – Update weights and biases
  – Compute the solution $u_{12,i}$ in the $\Omega_1 \cap \Omega_2$ with the respective Neural Networks $\mathcal{N}_i$
  – **if** $\|u_{12,1}^n - u_{12,2}^n\| \leq Tol$ **then**
    | STOP;
    end

end

Suppose we are only provided training data with each of the subdomains but no data in the overlap $\Omega_1 \cap \Omega_2$. This is shown in the left panel in Fig. 6 where the training data set is not complete. Using PINNS, the idea would be to solve for the whole domain by creating two different Neural Networks to solve u_i in $\Omega_i \setminus (\Omega_1 \cap \Omega_2)$ by PINNs. This is shown in the middle panel in Fig. 6. Employing only PINNS, we are able to generate a reasonable estimate of the exact solution with an L2-error of 0.42. Finally, to improve the accuracy, we employed coupled PINNs-DDM over the entire domain. This is shown in the last panel in Fig. 6. We noted that the coupled method is able to approximate the exact solution very well with an L2-error of 0.042. For generating these solutions 30,000 iterations for both PINNs and PINNs-DDM were employed.

Fig. 7 $\Omega = \Omega_1 \cup \Omega_2$

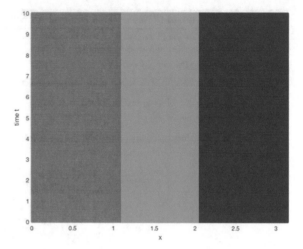

To show the application of PINNs-DDM algorithm created as a part of this work for multi-physics problems, we apply the method to the coupled system (2)–(3). Note that we use this system for simplicity of presentation, but one can extend this to other complex multi-physics systems as well. For this computation, we use

$$\alpha_V(x) = \frac{1}{20}, \quad \alpha_U(x) = \frac{1}{5}, \quad S_U = e^{-\frac{2}{5}t} \sin x \cos x, \quad S_V = 2e^{1-\frac{2}{5}t} \sin x \cos 2x$$

for $x \in [0, \pi]$ and $t \in [0, 10]$.

In order to validate our PINNs-DDM method, we evaluate the error against the exact solution for this system given by $V(x, t) = e^{1-\frac{2}{5}t} \sin 2x$ and $U(x, t) = e^{-\frac{2}{5}t} \sin x$.

For the computational domain shown in Fig. 7 where $\Omega = [0, \pi] \times [0, 10]$, we assume that we have data in sub-domain one (red rectangle) and in sub-domain 2 (blue rectangle), but no training data in the overlap (green rectangle). As prescribed by Algorithm 2, we create two Neural Networks that both have both spatial and temporal inputs (x, t), 3 hidden layers with size of 50 nodes and 2 outputs V and U that are coupled. Also, we choose randomly 500 training data points for each domain as well in the overlap, along with 500 input solutions at the points (x, t) for training.

Figure 8 shows the plots for errors between the PINNs-DDM solution and the exact solution in each sub-domain after 50, 000 iterations of training. The plots on the left denote the approximation for U and on the right for V. Clearly, the errors indicate superior performance of the PINNs-DDM method.

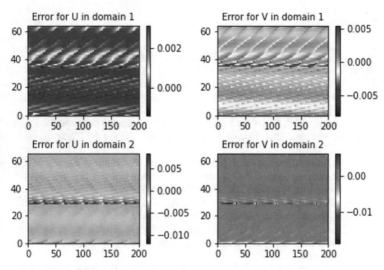

Fig. 8 Performance of PINNs-DDM for a coupled multi-physics problem

3 Discussion and Future Work

In this work, we combine a physics informed neural networks (PINNs) approach with a domain decomposition method (DDM) approach to yield a highly efficient methodology called PINNs-DDM for solving multi-physics problems is developed. While the examples and benchmark problems used in this work are not in higher dimensions, they provide the necessary insight into the efficiency of the method. One of the interesting findings from this work includes the performance of PINNs-DDM when only limited data is available to train. Our results suggest that PINNs-DDM is a robust candidate for solving complex system of PDEs motivated by real-world applications.

While this work has helped us to develop and design an efficient PINNs-DDM algorithm, there is still work that needs to be done to understand the convergence mathematically. As motivated in the introduction, the reason for exploring the system (2)–(3) is to understand the spread of droplet concentration (for example, from COVID-19) in the presence of airflow in enclosed spaces (such as aircraft cabins). Coupled with these are models that involve epidemiological equations that model the spread of a disease such as COVID-19. We hope to apply PINNs-DD to such an application in higher dimensions as well as equations that admit discontinuous coefficients α_V and α_U in a forthcoming paper. Another interesting aspect is to employ PINNs-DDM as an inverse approach to conduct parameter identification which is another aspect that will be investigated in the future.

Acknowledgements This work is supported in part by the Computational Mathematics program at the National Science Foundation through grant DMS 2031027 and DMS 2031029.

References

1. Aulisa E, Manservisi S, Seshaiyer P (2005) A non-conforming computational methodology for modeling coupled problems. Nonlinear Anal: Theory, Methods Appl 63(5–7):e1445–e1454
2. Aulisa E, Manservisi S, Seshaiyer P (2006) A computational multilevel approach for solving 2D Navier-Stokes equations over non-matching grids. Comput Methods Appl Mech Eng 195(33–36):4604–4616
3. Aulisa E, Manservisi S, Seshaiyer P (2008) A multilevel domain decomposition approach to solving coupled applications in computational fluid dynamics. Int J Numer Methods Fluids 56(8):1139–1145
4. Aulisa E, Cervone A, Manservisi S, Seshaiyer P (2009) A multilevel domain decomposition approach for studying coupled flow applications. Commun Comput Phys 6(2):319
5. Baker AJ, Taylor MB, Winowich NS, Heller MR (2000) Prediction of the distribution of indoor air quality and comfort in aircraft cabins using computational fluid dynamics (CFD). In: Air quality and comfort in airliner cabins, ASTM international
6. Bernardi C, Maday Y, Patera AT (1993) Domain decomposition by the mortar element method. In: Asymptotic and numerical methods for partial differential equations with critical parameters. Springer, Dordrecht, pp 269–286
7. Baydin AG, Pearlmutter BA, Radul AA, Siskind JM (2018) Automatic differentiation in machine learning: a survey. J Mach Learn Res 18
8. Cao Y, Gunzburger M, He X, Wang X (2014) Parallel, non-iterative, multi-physics domain decomposition methods for time-dependent Stokes-Darcy systems. Math Comput 83(288):1617–1644
9. Farhat C, Mandel J, Roux FX (1994) Optimal convergence properties of the FETI domain decomposition method. Comput Methods Appl Mech Eng 115(3–4):365–385
10. Gander MJ, Halpern L, Nataf F (2000) Optimized Schwarz methods. In: 12th international conference on domain decomposition methods, pp 15–27
11. Gander MJ, Hairer E (2008) Domain decomposition methods in science and engineering XVII
12. Hathway EA, Noakes CJ, Sleigh PA, Fletcher LA (2011) CFD simulation of airborne pathogen transport due to human activities. Build Environ 46(12):2500–2511
13. Kingma DP, Ba J (2014) Adam: a method for stochastic optimization. arXiv:1412.6980
14. Nielsen PV (1974) Flow in air conditioned rooms: model experiments and numerical solution of the flow equations
15. Lo LM (1997) Numerical studies of airflow movement and contaminant transport in hospital operating rooms. University of Minnesota, M.Sc. Thesis
16. LeCun Y, Bengio Y, Hinton G (2015) Deep learning. Nature 521(7553):436–444
17. Paszke A, Gross S, Chintala S, Chanan G, Yang E, DeVito Z, Lin Z, Desmaison A, Antiga L, Lerer A (2017) Automatic differentiation in Pytorch
18. Qi W, Seshaiyer P, Wang J (2021) A four-field mixed finite element method for Biot's consolidation problems. Electron Res Arch 29(3):2517
19. Qi W, Seshaiyer P, Wang J (2021) Finite element method with the total stress variable for Biot's consolidation model. Numer Methods PartL Differ Equ 37(3):2409–2428
20. Ramsundar B, Zadeh RB (2018) TensorFlow for deep learning: from linear regression to reinforcement learning. O'Reilly Media, Inc.
21. Raissi M, Seshaiyer P (2014) A multi-fidelity stochastic collocation method for parabolic partial differential equations with random input data. Int J Uncertain Quantif 4(3)
22. Raissi M, Seshaiyer P (2018) Application of local improvements to reduced-order models to sampling methods for nonlinear PDEs with noise. Int J Comput Math 95(5):870–880
23. Raissi M, Karniadakis GE (2018) Hidden physics models: machine learning of nonlinear partial differential equations. J Comput Phys 357:125–141
24. Raissi M, Perdikaris P, Karniadakis GE (2019) Physics-informed neural networks: a deep learning framework for solving forward and inverse problems involving nonlinear partial differential equations. J Comput Phys 378:686–707

25. Seshaiyer P, Smith P (2003) A non-conforming finite element method for sub-meshing. Appl Math Comput 139(1):85–100
26. Seshaiyer P, Suri M (2000) Uniform HP-convergence results for the mortar finite element method. Math Comput 69(230):521–546
27. Schwarz HA (1870) Ueber einen Grenzübergang durch alternirendes Verfahren. Zürcher u, Furrer
28. Shin Y, Darbon J, Karniadakis GE (2020) On the convergence of physics informed neural networks for linear second-order elliptic and parabolic type PDEs. arXiv:2004.01806
29. Smith BF (1997) Domain decomposition methods for partial differential equations. In: Parallel numerical algorithms. Springer, Dordrecht, pp 225–243
30. Swim EW, Seshaiyer P (2006) A nonconforming finite element method for fluid-structure interaction problems. Comput Methods Appl Mech Eng 195(17–18):2088–2099

Estimation of Current Earthquake Hazard Through Nowcasting Method

Sumanta Pasari⦿

Abstract In several tectonically active regions of the world, large magnitude earthquakes on fault systems are observed to occur in near-repetitive cycles as a consequence of stress accumulation and moment release. Since absolute measurements of stress–strain is unavailable through direct observations at all regions of interest, the area-based nowcasting method based on earthquake data is a potential alternative to estimate the uncertain current state of earthquake hazard in a defined region. Using the concept of natural-time counts, the nowcasting result comprises time-dependent earthquake potential score—a numerical quantification of earthquake-cycle progression since the last major event in the region. The nowcast score may be linked to the instantaneous risk of large events. This paper summarizes some basic formulation and key concepts of earthquake nowcasting with a demonstration of its applicability in disaster preparation and risk estimation. A case study from Java, Indonesia, is considered for illustration.

Keywords Earthquake nowcasting · Natural times · Hazard analysis

1 Introduction

Major earthquakes in large seismically active regions are often observed to occur in approximately repetitive cycles, though interoccurrence times or interevent small earthquake counts (natural times) exhibit randomness [1, 2]. Therefore, adequate knowledge of the current level of earthquake-cycle progression since the last major event in a geographic region is of great importance to many practical applications, such as policymaking, city planning, insurance, large-scale engineering constructions, and seismic awareness campaigns [1–5]. The earthquake nowcasting method [3–5] uses a simple concept of earthquake-cycle in the natural time domain [6], where the distribution of intermittent small earthquake counts (say, $M \geq 4.0$) between pairs of large events (say, $M >= 6.0$) carries significant information of the time-varying

S. Pasari (✉)
Department of Mathematics, Birla Institute of Technology and Science, Pilani Campus, Pilani, RJ 333031, India

© The Author(s), under exclusive license to Springer Nature Singapore Pte Ltd. 2022 55
R. Srinivas et al. (eds.), *Advances in Computational Modeling and Simulation*,
Lecture Notes in Mechanical Engineering,
https://doi.org/10.1007/978-981-16-7857-8_5

cycle progression in a network of geological faults [3–5]. As a consequence, earthquake potential scores (EPSs) of a number of local city regions embedded in a large spatial area allow comparing their current exposure to seismic hazard (risk) for disaster fund allocation and allied purposes. As the focus is to analyze the current (or, a very short-span of future) state of regional fault-system, the nowcasting method does not necessarily provide an implication of the long-term seismic behavior of the study region. Nonetheless, a deeper knowledge of the current state of hazard certainly helps in estimating the future state of earthquake hazard in a given region [3].

Conceptually, earthquake nowcasting method may be treated as a stochastic renewal process where the EPS increases upon every moment when the natural time increments by unity. By definition of the renewable process, history before the last large earthquake is all forgotten, though it remembers what happened after the last major event through the current number of small event counts $n(t)$ at the time of evaluation t. The nowcasting idea essentially reflects a short-term fault memory in which the system has no leftover stress for the next cycle, unlike the idea of earthquake supercycle in tectonics [1–5]. With the onset of a seismic cycle, the process starts, continues and the nowcast score increases with the occurrence of each small magnitude event. The EPS at a specific time can be high, and it will remain high until a nucleation event (major earthquake) occurs to tip the system over into a new potential well. Thus, nowcast scores provide direct relevance to estimating seismic risk of a study region as long as natural time seismicity statistics does not exhibit a stationary Poisson process. In the latter case, the nowcasting method still provides a valid measure of earthquake cycle progression, though the estimation of seismic risk (conditioned on the elapsed time, say $n(t)$, since the last major event) turns out to be irrelevant to the time-dependent EPS [3–5]. Therefore, in general, EPS determines the youngness or oldness of earthquake cycle through the observed age of $n(t)$ in a nice 0–100% scale of extremeness.

The seismic nowcasting approach has been applied to several regional and global megacities of United States, Chile, Taiwan, Philippines, Indonesia, India, Nepal, Pakistan, and Bangladesh to obtain a snapshot evaluation of the current progress of hazard cycle in the respective local regions [3–15]. The method was demonstrated for a few locations of induced seismicity, great-earthquakes, and mega-tsunamis globally [3, 4, 9]. Attempts were made to study possible interconnections among earthquake nowcasting, earthquake forecasting, and their relevance to risk estimation as well as stress–strain distribution [7].

This article summarizes some basic formulation and key concepts of earthquake nowcasting with an emphasis on the mathematical explanation of its relevance to risk estimation. A case study from the seismically active Java Island, Indonesia, has been considered for illustration.

2 Formulation

In a large network of fault system, earthquake event occurrences are often associated with a frequency-magnitude scaling relationship [1]. Thus, the ensemble statistics of natural-times, interspersed small-magnitude events (say, M >= 4.0) between successive large events (say, M >= 6.0), may prove to be useful in hazard analysis. The nowcasting method basically uses a counting of discrete events to mark the development of the system, rather than the usual consideration of clock or calendar time. This area-based approach requires a defined geographic region, just as the fault-based approaches require definition of the target faults [3]. Comparison of EPS as time progresses in these defined regions is justified as long as the defined region remains constant. The key task in nowcasting analysis is to develop seismicity statistics from sufficient number of natural times so that the developed statistics can be utilized to compute EPS in terms of the cumulative distribution function (CDF) evaluated at the current count of small events $n(t)$. A number of reference probability distributions may be used for this purpose.

Let $C_1, C_2, C_3, \cdots, C_k$ be k number of local city regions within a larger seismic region C. These local city regions are often described by the area of radius R (say, 300 km) from their respective city centers. In the earthquake catalog of region C, let there be m cycles of large events and let $\{N_1, N_2, \cdots, N_m\}$ be the corresponding random sample of natural times (N). From the data-derived distribution of N and known small event count $n(t)$ of a local region at the time of evaluation t, the EPS for the local region is defined as $EPS(n(t)) \equiv F_N(n(t)) = P\{N \le n(t)\}$. Notice that EPS is an increasing function of time (natural-time) and it provides a measure of cycle progression since the last major event in the local region, though its relevance to seismic hazard (risk) is a separate problem.

In an ongoing seismic cycle, the current level of seismic hazard (instantaneous risk) can be assessed through conditional probability of a large earthquake in the near future (in natural-time domain) on the basis of the current state $EPS(n(t))$. In this case, one can formulate a definition of extremely short-term risk, P_{next}^{large}, which is the probability that the next earthquake in the local region is actually a large event. In fact, to observe the risk of large events in the entire cycle, one has to derive the probability distribution of the waiting time τ until the next large earthquake in the local region. At this point, it may be emphasized that EPS through CDF or survival function describes the distribution of the "waiting time between successive large earthquakes", while the risk assessment concerns "the waiting time until the next large earthquake" based on the current status $EPS(n(t))$.

3 Illustration of Nowcasting Method

To illustrate the method of seismic nowcasting, an earthquake catalog (1963–2021) of Java Island, Indonesia, is considered (Fig. 1). Based on the magnitude-completeness

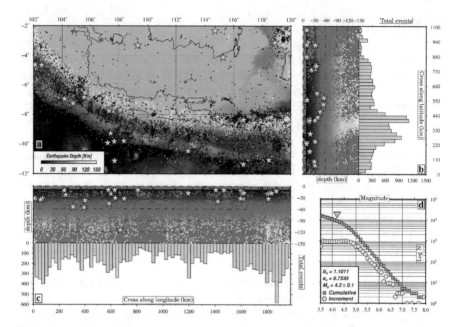

Fig. 1 Epicentral locations of earthquake data (1963–2020) on Java Island **a**; subplots **b**–**d** show cross-section maps of the focal depth and magnitude completeness value [12]

of this catalog, the threshold of "small" events turns out to be $M_\sigma = 4.0$, while the threshold of "large" events is decided as $M_\lambda = 6.5$ [12]. There are 31 cycles of large events, providing a random sample of size 31 to develop natural-time statistics of the region. Within the large study region, two local city regions, namely Surabaya and Jakarta, are considered with their respective areas defined by radius $R = 300$ km [12].

Three reference distributions, namely exponential, gamma, and Weibull are considered to fit observed natural-time counts [16–22]. Parameters are estimated from the maximum likelihood method, and the relative suitability of studied distributions is assessed through the non-parametric Kolmogorov–Smirnov criterion [16, 17]. It is found that the Weibull distribution has the most suitable representation, meaning that the study region reveals a natural time Weibull statistics. Using the current counts of small events $n(t) = 829$ and $n(t) = 127$, the nowcast scores of Surabaya and Jakarta are 89% and 43%, respectively [12]. This means that Surabaya has progressed about 89% of its earthquake cycle of large events, whereas Jakarta has progressed about 43% of its cycle. With this information, the instantaneous seismic risk may be computed for policymaking and city planning.

Usually, a sensitivity study of the threshold values (e.g., city radius, small and large magnitudes, and city region) is associated with earthquake nowcasting [3–5], though it has been deliberately skipped in the present manuscript.

4 Summary

Seismic nowcasting is an area-based indirect method to evaluate the current progression of earthquake cycle of large events. It helps to assess the present state of regional fault system and thereby provides some useful information to the city-planners, engineers, and entrepreneurs. Basic formulation of the nowcasting method and some key concepts are discussed with an emphasis on its relevance to risk estimation. The case study from Java, Indonesia, has demonstrated the applicability of earthquake nowcasting in hazard management as well as planning and design considerations. As on 18 February, 2021, Surabaya observes an EPS score of 89%, while it is 43% for Jakarta.

The earthquake nowcasting that mainly refers to the estimation of the current state of earthquake hazard can also be used in earthquake forecasting or plate tectonic studies using natural time Weibull projection, slider-block toy model, and space–time clustering behavior of bursts of small earthquakes in seismogenic areas [3, 7]. Like conventional fault-based earthquake hazard analysis, the area-based nowcasting approach offers a unique characterization of the spatial distribution of regional earthquake hazard in terms of earthquake potential score at selected regions. The methodology discussed here is general, scalable, and transportable to any seismic area and in essence, it may serve as a decision support system for disaster preparation and risk estimation.

Acknowledgements I thank the editor and two anonymous reviewers for their extremely careful and thoughtful reviews. The Generic Mapping Tool (GMT) was used to prepare the figure. Earthquake data were downloaded from Advanced National Seismic System comprehensive catalog and International Seismological Centre catalog. The financial supports from DST-FIST grant to the Department of Mathematics, BITS Pilani is duly acknowledged. Partial financial support was available from a start-up research grant of IRDR-ICoE Taipei, Taiwan.

References

1. Scholz CH (2019) The mechanics of earthquakes and faulting. Cambridge University, Cambridge
2. Pasari S (2015) Understanding Himalayan tectonics from geodetic and stochastic modeling. PhD thesis, Indian Institute of Technology Kanpur, India, pp 1–376
3. Rundle JB, Turcotte DL, Donnellan A, Grant-Ludwig L, Luginbuhl M, Gong G (2016) Nowcasting earthquakes. Earth Space Sci 3:480–486
4. Rundle JB, Luginbuhl M, Khapikova P, Turcotte DL, Donnellan A, McKim G (2020) Nowcasting great global earthquake and tsunami sources. Pure Appl Geophys 177:359–368
5. Pasari S (2019) Nowcasting earthquakes in the Bay of Bengal region. Pure Appl Geophys 176(4):1417–1432
6. Varotsos PA, Sarlis NV, Skordas ES (2011) Natural time analysis: the new view of time. Springer, Berlin
7. Rundle JB, Donnellan A (2020) Nowcasting earthquakes in southern California with machine learning: bursts, swarms and aftershocks may reveal the regional tectonic stress. Earth Space Sci 58:147

8. Pasari S, Sharma Y (2020) Contemporary earthquake hazards in the west-northwest Himalaya: a statistical perspective through natural times. Seismol Res Lett 91(6):3358–3369
9. Luginbuhl M, Rundle JB, Turcotte DL (2018) Natural time and nowcasting induced seismicity at the Groningen gas fields in the Netharlands. Geophys J Int 215:753–759
10. Pasari S, Mehta A (2018) Nowcasting earthquakes in the northwest Himalaya and surrounding regions. ISPRS-Int Arc Photogram Rem Sens Appl 42(5):855–859
11. Pasari S (2020) Stochastic modeling of earthquake interevent counts (natural times) in northwest Himalaya and adjoining regions. In Bhattacharyya S, Kumar J, Ghoshal K (eds) Mathematical modeling and computational tools. Springer Proceedings in Mathematics & Statistics, vol 320, pp 495–501
12. Pasari S, Sharma Y, Neha (2021) Quantifying the current state of earthquake hazards in Nepal. Appl Comput Geosci 10:100058 (2021)
13. Pasari S, Simanjuntak AVH, Mehta A, Neha, Sharma Y (2021) The current state of earthquake potential on Java Island, Indonesia. Pure Appl Geophys (2021, in print), vol 178, pp 2789–2806 (2021)
14. Pasari S, Simanjuntak AVH, Mehta A, Neha, Sharma Y (2021) A synoptic view of the natural time distribution and contemporary earthquake hazards in Sumatra, Indonesia. Nat Hazards 108(1):309–321
15. Pasari S, Simanjuntak AVH, Neha, Sharma Y Nowcasting earthquakes in Sulawesi Island, Indonesia. Geosci Lett 8(27):1–13
16. Pasari S, Dikshit O (2014) Three-parameter generalized exponential distribution in earthquake recurrence interval estimation. Nat Hazards 73:639–656
17. Pasari S, Dikshit O (2014) Impact of three-parameter Weibull models in probabilistic assessment of earthquake hazards. Pure Appl Geophys 171(7):1251–1281
18. Pasari S, Dikshit O (2015) Distribution of earthquake interevent times in northeast India and adjoining regions. Pure Appl Geophys 172(10):2533–2544
19. Pasari S, Dikshit O (2015) Earthquake interevent time distribution in Kachchh, Northwestern India. Earth, Planets Space 67:129
20. Pasari S, Dikshit O (2018) Stochastic earthquake interevent time modelling from exponentiated Weibull distributions. Nat Hazards 90(2):823–842
21. Pasari S (2018) Stochastic modeling of earthquake interoccurrence times in northwest Himalaya and adjoining regions. Geomat, Nat Hazards Risk 9(1):568–588
22. Pasari S (2019) Inverse Gaussian versus lognormal distribution in earthquake forecasting: keys and clues. J Seism 23(3):537–559

Global Uniqueness Theorem
for a Discrete Population Balance Model
with Application in Astrophysics

Sonali Kaushik and Rajesh Kumar

Abstract The article presents the uniqueness results for the discrete Oort–Hulst–Safronov equation for the unbounded product coagulation kernel $C_{j,k} \le jk \; \forall \; j, k \in \mathbb{N}$. The contraction mapping theorem is used along with the properties of the second and third moments in proving that the solution is unique.

Keywords Oort–Hulst–Safronov (OHS) model · Discrete population balance equation · Product kernel · Moments · Uniqueness

1 Introduction

The discrete OHS or Safronov–Dubovski coagulation equation was introduced by Dubovski [1, 2] in 1999 and then further studied by Bagland [3] and Davidson [4, 5]. The equation is given by

$$\frac{d\phi_j(\tau)}{d\tau} = \phi_{j-1}(\tau) \sum_{k=1}^{j-1} kC_{j-1,k}\phi_k(\tau) - \phi_j(\tau) \sum_{k=1}^{j} kC_{j,k}\phi_k(\tau) - \sum_{k=j}^{\infty} C_{j,k}\phi_j(\tau)\phi_k(\tau), \quad (1)$$

for $\tau \in [0, \infty)$, $j, k \in \mathbb{N}$ and with the initial condition

$$\phi_j(0) = \phi_j^{in}. \quad (2)$$

The equation explains a process of collision between particles of mass js_0 and ks_0, where $s_0 > 0$ is the mass of the smallest particle in the system. This leads to the

---

S. Kaushik · R. Kumar (✉)
Department of Mathematics, Birla Institute of Technology & Science Pilani, Pilani Campus, Pilani, RJ, India
e-mail: rajesh.kumar@pilani.bits-pilani.ac.in

S. Kaushik
e-mail: p20180023@pilani.bits-pilani.ac.in

© The Author(s), under exclusive license to Springer Nature Singapore Pte Ltd. 2022
R. Srinivas et al. (eds.), *Advances in Computational Modeling and Simulation*,
Lecture Notes in Mechanical Engineering,
https://doi.org/10.1007/978-981-16-7857-8_6

formation of monomers from the x-mer (a particle of mass $x s_0$), which, in turn, results in the coalescence of the monomers with the y-mer, where $x \leq y$ (here $x, y = j, k$). The first term depicts the addition of a j sized cluster in the system due to the coagulation of an $(j - 1)$-mer and a monomer. The second and third terms followed by a negative sign signifies the death of a j-mer. The second expression is present due to the formation of particles of mass greater than $j s_0$. The third term is the flag-bearer for the case $x = j$ and $y = k$, i.e., when a j-mer breaks into monomers and they coagulate with a k-mer. Here, $\phi_j(\tau)$ is the concentration of particles of size j at time τ and $C_{j,k}$, $j \neq k$ is the rate at which particles of sizes j and k collide and is called the coagulation kernel. Moreover, $C_{j,j}$ can be calculated by halving the collision rate of a j-mer. The kernel $C_{j,k}$ is assumed to be non-negative and symmetric, i.e., $C_{j,k} \geq 0$ and $C_{j,k} = C_{k,j}$.

In addition to the concentration of particles, following moments are also of interest in the sense of using in real life applications or in proving some theoretical results. The r_{th} moment is given by

$$M_r(\tau) = \sum_{j=1}^{\infty} j^r \phi_j(\tau), \tag{3}$$

in which $r = 0$ leads to the total number of particles (zeroth moment) and the total mass in the system (first moment) is obtained by taking $r = 1$.

The expression involves two major events, collision followed by coagulation. There are various examples of the phenomenon explained by the Safronov–Dubovksi equation (being the discrete version of the Oort–Hulst–Safronov equation [6, 7]), namely, the collision of asteroids [8], formation of saturn's rings [9], formation of protoplanetary disc around a newly formed star [10]. There are a number of real life problems caused due to the coagulation of particles, be it the health hazards of sand and dust storms [11], analysis of nano-metal dust explosions [12] or the impact on glacier mass as an aftermath of the accumulation of volcanic dust [13].

Very few literatures are available on the well-posedness of such model and also in the direction of numerical implementation. Bagland [3] considered the Eq. (1) and discussed the existence of solution for a bounded kernel of the form,

$$\lim_{l \to \infty} \frac{C_{j,l}}{l} = 0$$

with the initial condition satisfying $\sum_{j=1}^{\infty} j \phi_j^{in}(0) < \infty$. In 2014, Davidson [4] proved the global existence of the solution for bounded as well as unbounded kernels. The results are shown for the bounded kernel of the form $k C_{j,k} \leq M$ when $k \leq j$ and M being a positive constant while the unbounded kernels $C_{j,k} \leq M g_j g_k$ are taken under the assumption that $\frac{g_j}{j} \to 0$ as $j \to \infty$. His work includes the uniqueness result but only for the bounded kernel $C_{j,k} \leq M$. Further, in his dissertation [5], the existence theory is investigated for product kernel $C_{j,k} \leq c_1 j k$, where c_1 is a constant. The solutions are proven to be non-negative and exist in the space $l_{1,j}$ with the norm

$$||z||_1 = \sum_{j=1}^{\infty} j|z_j|.$$

The assumptions used to prove the existence include the initial second moment to be finite and the bound for the second moment is finite, i.e.,

$$M_2(0) < \infty \quad \text{and} \quad \sup_{0 \leq \tau \leq T} M_2(\tau) = N_2 < \infty.$$

The author [5] has also included the plots of $M_0(t)$, $M_1(t)$ and $M_2(t)$ for $t = 0, 0.1, 0.2$ and plotted the numerical solutions using an implicit iteration scheme given by

$$\frac{\phi_j^{(n+1)} - \phi_j^{(n)}}{\eta} = \phi_{j-1}^{(n+1)}(\tau) \sum_{k=1}^{j-1} jk^2 \phi_k^{(n+1)}(\tau) - \phi_j^{(n+1)}(\tau) \sum_{k=1}^{j} jk^2 \phi_k^{(n+1)}(\tau) - \sum_{k=j}^{J} jk \phi_j^{(n+1)}(\tau) \phi_k^{(n)}(\tau) \quad (4)$$

with

$$\phi_j(0) = \frac{1}{2^j} \quad \text{for} \quad j \geq 1. \tag{5}$$

Here, the time step $\eta = 0.0025$ and the maximum value of j are taken as $J = 5000$. But, the comparison of numerical solutions with the analytical solutions was not carried out. For this, we need the guarantee that there is only one analytical solution so that the error calculation can be validated. This requires the study of uniqueness theory for the Eq. (1). So far, such results are not available for the product kernel. Therefore, our main aim in this article is to deal with the analysis of conditions under which the solution to the Eq. (1) is unique when $C_{j,k} \leq jk$ for all $j, k \geq 1$.

The contraction mapping theorem and the Gronwall's lemma play a vital role in proving the uniqueness result. The bounds of higher moments, in particular, up to third moment was indispensable for the application of the theorem. To analyze our results, the weighted space $l_{1,j}$ is considered with norms defined by

$$||z||_1 = \sum_{j=1}^{\infty} j^2 |z_j| \tag{6}$$

and

$$|||z||| = \sup_{0 \leq \tau \leq T} ||z||_1. \tag{7}$$

From (1), the definition of the solution $\phi_j(\tau)$ can be given as

$$\phi_j(\tau) = \phi_j(0)$$

$$+ \int_0^\tau \begin{pmatrix} \phi_{j-1}(s) \sum_{k=1}^{j-1} k C_{j-1,k} \phi_k(s) \\ -\phi_j(s) \sum_{k=1}^{j} k C_{j,k} \phi_k(s) \\ -\sum_{k=j}^{\infty} C_{j,k} \phi_j(s) \phi_k(s) \end{pmatrix} ds := L(\phi)_j \qquad (8)$$

where L is the non-linear operator. The second and third moments are of special interest here and are, respectively, defined as

$$M_2(\tau) = \sum_{j=1}^{\infty} j^2 \phi_j(\tau) \qquad (9)$$

and

$$M_3(\tau) = \sum_{j=1}^{\infty} j^3 \phi_j(\tau). \qquad (10)$$

The paper is organized as follows. In Sect. 2, the lemma providing the bound for the third moment is stated and proved. The result established in the lemma is required to prove the uniqueness result, which is further presented in Sect. 3. The Sect. 4 includes some numerical results regarding the product kernel. Finally, some conclusions are placed at the end of the article.

2 Bounds of Moments

Here, we analyze the third moment and discuss its boundedness over a finite time interval. The proof makes use of a property of the first moment, namely $M_1(\tau) \leq M_1(0)$; see [4]. Another assumption required to establish the result is the boundedness of the second moment, same as in [5], i.e.

$$\sup_{0 \leq \tau \leq T} \sum_{j=1}^{n} j^2 \phi_j(\tau) = \alpha \qquad (11)$$

where α is a constant depending on T.

Lemma 2.1 *Let $\phi_j(\tau)$ is the solution of (1) such that the Eq. (11) holds and $M_i(0)$, $i = 1, 2, 3$ are finite. Then the third moment $M_3(\tau)$, defined by (10) is bounded for the product kernel $C_{j,k} \leq jk$ for all $j, k \geq 1$ such as*

$$M_3(\tau) \le \beta \tag{12}$$

when $\tau \in [0, T]$ and β is a constant.

Proof Multiplying the Eq. (8) by j^3 on both the sides and taking summation over j from 1 to n lead to

$$\sum_{j=1}^{n} j^3 \phi_j(\tau) = \sum_{j=1}^{n} j^3 \phi_j(0)$$

$$+ \int_0^\tau \left(\begin{array}{c} \sum_{j=2}^{n} j^3 \phi_{j-1} \sum_{k=1}^{j-1} kC_{j-1,k}\phi_k \\ -\sum_{j=1}^{n} j^3 \phi_j \sum_{k=1}^{j} kC_{j,k}\phi_k \\ -\sum_{j=1}^{n} j^3 \phi_j \sum_{k=j}^{\infty} C_{j,k}\phi_k \end{array} \right)(s)ds.$$

Substituting $j - 1$ by j', further taking $j' \to j$ in the first term of the above equation and some simplifications yield

$$\sum_{j=1}^{n} j^3 \phi_j(\tau) = \sum_{j=1}^{n} j^3 \phi_j(0)$$

$$+ \int_0^\tau \left(\begin{array}{c} \sum_{j=1}^{n} \sum_{k=1}^{j} (3j^2 + 3j + 1)kC_{j,k}\phi_j(s)\phi_k(s) \\ -(n+1)^3 \phi_n(s) \sum_{k=1}^{n} kC_{n,k}\phi_k(s) \end{array} \right)$$

$$- \int_0^\tau \left(\sum_{j=1}^{n} \sum_{k=j}^{\infty} j^3 \phi_j(s) C_{j,k}\phi_k(s) \right)$$

$$\le \sum_{j=1}^{n} j^3 \phi_j(0) + \int_0^\tau \sum_{j=1}^{n} \sum_{k=1}^{j} (3j^2 + 3j + 1)kC_{j,k}\phi_j(s)\phi_k(s)ds.$$

Now, putting $C_{j,k} \le jk$ and using (11) one can obtain for $\tau \in [0, T]$

$$\sum_{j=1}^{n} j^3 \phi_j(\tau) \le \sum_{j=1}^{n} j^3 \phi_j(0) + \int_0^\tau 3\alpha \sum_{j=1}^{n} j^3 \phi_j(s)ds + 3\alpha^2 T + \alpha M_1(0)T.$$

Thus, Gronwall's lemma concludes that

$$\sum_{j=1}^{n} j^3 \phi_j(\tau) \leq (M_3(0) + 3\alpha^2 T + \alpha M_1(0)T)e^{3\alpha\tau}.$$

The finiteness of $M_3(0)$, α and $M_1(0)$ allows us to pass the limit into the equation and so letting $n \to \infty$ gives the above expression as

$$M_3(\tau) \leq \beta, \quad \tau \in [0, T]$$

for β is a constant depending on $M_3(0)$, α, $M_1(0)$ and T. $\qquad\square$

3 Uniqueness

Theorem 3.1 *Let $\phi_j(\tau)$ be a non-negative solution of the Eq. (1) and α, β are the constants defined earlier. Also, the following hypotheses*

1. *(H1)* $\|\phi\|_1 \geq 1$ *and*
2. *(H2)* $\frac{5\beta + 7\alpha}{12\alpha(\beta+1)} < 1$
   hold, then $\phi_j(\tau)$ is the only solution on $[0, \infty)$.

Proof The outline of the proof is as follows. First, the bound for T is calculated which will be helpful in proving that there is a contraction mapping for $L : l_{1,j} \to l_{1,j}$ endowed with $||| \cdot |||$ norm. This step proves the continuity of the solution $\phi_j(\tau)$. Then, a function $u(\tau)$ is defined as the difference between the two solutions of the Safronov–Dubovksi equation which ensures the non-negativity of $u(\tau)$. Finally, $u(\tau)$ is proven to be identically zero by the application of Gronwall's lemma.
To find the bound, we assume that $||| \cdot |||$ satisfies the invariance relation

$$|||L(\phi)||| \leq |||\phi|||. \tag{13}$$

Thus, by the definition of norm (6), it yields

$$\|L(\phi)\|_1 = \sum_{j=1}^{\infty} j^2$$

$$\left| \phi_j(0) + \int_0^\tau \left(\phi_{j-1}(s) \sum_{k=1}^{j-1} kC_{j-1,k}\phi_k(s) - \phi_j(s) \sum_{k=1}^{j} kC_{j,k}\phi_k(s) - \sum_{k=j}^{\infty} C_{j,k}\phi_j(s)\phi_k(s) \right) ds \right|.$$

Replacing $j - 1$ by j' and then $j' \leftrightarrow j$ in the above expression yields

$$\|L(\phi)\|_1 \leq M_2(0) + \sum_{j=1}^{\infty} 2j^2 \left| \int_0^{\tau} \phi_j(s) \sum_{k=1}^{j} kC_{j,k}\phi_k(s)ds \right| + \sum_{j=1}^{\infty} 2j \left| \int_0^{\tau} \sum_{k=1}^{j} kC_{j,k}\phi_k(s)\phi_j(s)ds \right|$$

$$+ \sum_{j=1}^{\infty} \left| \int_0^{\tau} \sum_{k=1}^{j} kC_{j,k}\phi_k(s)\phi_j(s)ds \right|$$

$$+ \sum_{j=1}^{\infty} j^2 \left| \int_0^{\tau} \sum_{k=j}^{\infty} C_{j,k}\phi_j(s)\phi_k(s)ds \right|.$$

Putting $C_{j,k} \leq jk \; \forall \; j,k \geq 1$, one can obtain

$$\|L(\phi)\|_1$$

$$\leq M_2(0) + \int_0^{\tau} \sum_{j=1}^{\infty} \sum_{k=1}^{j} \left(2j^3k^2 + 2j^2k^2 + jk^2 \right) |\phi_j(s)\phi_k(s)| ds$$

$$+ \int_0^{\tau} \sum_{j=1}^{\infty} \sum_{k=j}^{\infty} j^3 k |\phi_j(s)\phi_k(s)| ds.$$

Some simplifications and applying Lemma 2.1 lead to the following equation

$$\|L(\phi)\|_1 \leq \alpha + \int_0^{\tau} 2\beta\|\phi\|_1 + 2\|\phi\|_1^2 + \|\phi\|_1^2 + \beta\|\phi\|_1 ds.$$

Further, $(H1)$ and the definition of $||| \cdot |||$ guarantees that

$$|||L(\phi)||| \leq \alpha + 3|||\phi|||^2 T(\beta + 1).$$

For the invariance property to hold, the expression given below must hold true

$$\alpha + 3|||\phi|||^2 T(\beta + 1) - |||\phi||| \leq 0.$$

To have the real roots for the above inequality, the discriminant must be positive, which gives us the bound for T as

$$T \leq \frac{1}{12\alpha(\beta + 1)}. \tag{14}$$

Next, we need to show that, there is a contraction mapping for the non-linear operator L which is possible if

$$|||L(\phi) - L(\lambda)||| \leq b \; |||\phi - \lambda|||, \quad b < 1 \tag{15}$$

is proved. The definition of $L(\phi)$ yields

$$\|L(\phi) - L(\lambda)\|_1$$

$$= \sum_{j=1}^{\infty} j^2 \left| \phi_j(0) + \int_0^\tau \left(\phi_{j-1}(s) \sum_{k=1}^{j-1} kC_{j-1,k}\phi_k(s) - \phi_j(s) \sum_{k=1}^{j} kC_{j,k}\phi_k(s) - \sum_{k=j}^{\infty} C_{j,k}\phi_j(s)\phi_k(s) \right) ds \right.$$

$$\left. - \lambda_j(0) - \int_0^\tau \left(\lambda_{j-1}(s) \sum_{k=1}^{j-1} kC_{j-1,k}\lambda_k(s) - \lambda_j(s) \sum_{k=1}^{j} kC_{j,k}\lambda_k(s) - \sum_{k=j}^{\infty} C_{j,k}\lambda_j(s)\lambda_k(s) \right) ds \right|$$

$$\leq \sum_{j=1}^{\infty} (2j^2 + 2j + 1) \left| \int_0^\tau \sum_{k=1}^{j} kC_{j,k}(\phi_j(s)\phi_k(s) - \lambda_j(s)\lambda_k(s)) ds \right|$$

$$+ \sum_{j=1}^{\infty} j^2 \left| \int_0^\tau \sum_{k=j}^{\infty} C_{j,k}(\phi_j(s)\phi_k(s) - \lambda_j(s)\lambda_k(s)) ds \right|.$$

Further, by using the relation $xy - \tilde{x}\tilde{y} = \frac{1}{2}[(x-y)(\tilde{x}+\tilde{y}) + (x+y)(\tilde{x}-\tilde{y})]$ and $\phi_j(0) = \lambda_j(0)$, the above expression reduces to

$$\|L(\phi) - L(\lambda)\|_1 \leq \sum_{j=1}^{\infty} \left(j^2 + j + \frac{1}{2} \right)$$

$$\int_0^\tau \sum_{k=1}^{j} kC_{j,k}(|\phi_j(s) - \lambda_j(s)||\phi_k(s) + \lambda_k(s)| + |\phi_k(s) - \lambda_k(s)||\phi_j(s) + \lambda_j(s)|) ds$$

$$+ \sum_{j=1}^{\infty} \frac{j^2}{2} \int_0^\tau \sum_{k=j}^{\infty} C_{j,k}(|\phi_j(s) - \lambda_j(s)||\phi_k(s) + \lambda_k(s)| + |\phi_k(s) - \lambda_k(s)||\phi_j(s) + \lambda_j(s)|) ds$$

$$:= S_1 + S_2 + S_3 + S_4. \tag{16}$$

We simplify each term S_i, $i = 1, 2, 3, 4$ separately. Putting $C_{j,k} \leq jk$ and some computations, changing the order of summation, enable us to have

$$S_1 \leq \int_0^\tau \sum_{j=1}^{\infty} \sum_{k=j}^{\infty} j^2 k^3 |\phi_j(s) - \lambda_j(s)||\phi_k(s) + \lambda_k(s)| ds$$

$$+ \int_0^\tau \sum_{j=1}^{\infty} \sum_{k=1}^{j} j^3 k^2 |\phi_k(s) - \lambda_k(s)||\phi_j(s) + \lambda_j(s)| ds$$

$$\leq \int_0^\tau 4\beta \|\phi - \lambda\|_1 ds. \tag{17}$$

Secondly, by simplifying S_2, one can easily obtain

$$S_2 \leq \int_0^\tau \sum_{j=1}^{\infty} \sum_{k=1}^{j} j^2 k^2 (|\phi_j(s) - \lambda_j(s)||\phi_k(s) + \lambda_k(s)| + |\phi_k(s) - \lambda_k(s)||\phi_j(s) + \lambda_j(s)|) ds$$

$$\leq \int_0^\tau 4\alpha \|\phi - \lambda\|_1 ds. \tag{18}$$

The third sum from (16) can be evaluated as

$$S_3 \leq \int_0^\tau \sum_{j=1}^\infty \sum_{k=1}^j \frac{jk^2}{2} (|\phi_j(s) - \lambda_j(s)||\phi_k(s) + \lambda_k(s)| + |\phi_k(s) - \lambda_k(s)||\phi_j(s) + \lambda_j(s)|)ds$$

$$\leq \int_0^\tau \sum_{j=1}^\infty \sum_{k=1}^j \frac{j^2k^2}{2} (|\phi_j(s) - \lambda_j(s)||\phi_k(s) + \lambda_k(s)| + |\phi_k(s) - \lambda_k(s)||\phi_j(s) + \lambda_j(s)|)ds$$

$$\leq \int_0^\tau 2\alpha ||\phi - \lambda||_1 ds. \tag{19}$$

Finally, the term S_4 yields

$$S_4 \leq \int_0^\tau \sum_{j=1}^\infty \sum_{k=j}^\infty \frac{j^2k^2}{2} |\phi_j(s) - \lambda_j(s)||\phi_k(s) + \lambda_k(s)|ds$$

$$+ \int_0^\tau \sum_{j=1}^\infty \sum_{k=j}^\infty \frac{j^3k^2}{2} |\phi_k(s) - \lambda_k(s)||\phi_j(s) + \lambda_j(s)|ds$$

$$\leq \int_0^\tau (\alpha + \beta)||\phi - \lambda||_1 ds. \tag{20}$$

Hence, using the Eqs. (17)–(20) in (16), it is easy to see that

$$||L(\phi) - L(\lambda)||_1 \leq \int_0^\tau (5\beta + 7\alpha)||\phi - \lambda||_1 ds$$

which can further be taken as

$$|||L(\phi) - L(\lambda)||| \leq (5\beta + 7\alpha)|||\phi - \lambda|||T.$$

Using Eq. (14) and $(H2)$, the desired result is accomplished for $\tau \in [0, T]$. Thus, the solution $\phi_j(\tau)$ is continuous when $0 \leq \tau \leq T$ where $T < \frac{1}{12\alpha(\beta+1)}$. To proceed further, in order to achieve the uniqueness result, let us assume that $\phi_j(\tau)$ and $\lambda_j(\tau)$ are two solutions of the Safronov–Dubovski coagulation equation with the same initial condition, i.e.

$$\phi_j(0) = \lambda_j(0). \tag{21}$$

Now, consider

$$u(\tau) = \sum_{j=1}^\infty |\phi_j(\tau) - \lambda_j(\tau)|,$$

where $\phi_j(\tau)$ is given by the Eq. (8) and so we have

$$u(\tau) = \sum_{j=1}^{\infty} \left| \begin{array}{l} \phi_j(0) \\ + \int_0^{\tau} \left(\begin{array}{l} \phi_{j-1}(s) \sum_{k=1}^{j-1} kC_{j-1,k}\phi_k(s) - \phi_j(s) \sum_{k=1}^{j} kC_{j,k}\phi_k(s) \\ - \sum_{k=j}^{\infty} C_{j,k}\phi_j(s)\phi_k(s) \end{array} \right) ds - \lambda_j(0) \\ - \int_0^{\tau} \left(\begin{array}{l} \lambda_{j-1}(s) \sum_{k=1}^{j-1} kC_{j-1,k}\lambda_k(s) - \lambda_j(s) \sum_{k=1}^{j} kC_{j,k}\lambda_k(s) \\ - \sum_{k=j}^{\infty} C_{j,k}\lambda_j(s)\lambda_k(s) \end{array} \right) ds \end{array} \right| .$$

Using Eq. (21) and replacing $j - 1$ by j' in the first sum and fourth sum, the above equation becomes

$$u(\tau) \leq 2\sum_{j=1}^{\infty} \left| \int_0^{\tau} \sum_{k=1}^{j} kC_{j,k}(\phi_j(s)\phi_k(s) - \lambda_j(s)\lambda_k(s))ds \right| + \sum_{j=1}^{\infty} \left| \int_0^{\tau} \sum_{k=j}^{\infty} C_{j,k}(\lambda_j(s)\lambda_k(s) - \phi_j(s)\phi_k(s))ds \right|$$

$$\leq \int_0^{\tau} \sum_{j=1}^{\infty}\sum_{k=1}^{j} kC_{j,k}(|\lambda_j(s) - \phi_j(s)||\lambda_k(s) + \phi_k(s)| + |\lambda_j(s) + \phi_j(s)||\lambda_k(s) - \phi_k(s)|)ds$$

$$+ \int_0^{\tau} \sum_{j=1}^{\infty}\sum_{k=j}^{\infty} \frac{C_{j,k}}{2}(|\lambda_j(s) - \phi_j(s)||\lambda_k(s) + \phi_k(s)| + |\lambda_j(s) + \phi_j(s)||\lambda_k(s) - \phi_k(s)|)ds.$$

Now, substituting $C_{j,k} \leq jk$ yields

$$u(\tau) \leq \int_0^{\tau} \sum_{j=1}^{\infty}\sum_{k=1}^{j} jk^2(|\lambda_j(s) - \phi_j(s)||\lambda_k(s) + \phi_k(s)| + |\lambda_j(s) + \phi_j(s)||\lambda_k(s) - \phi_k(s)|)ds$$

$$+ \int_0^{\tau} \sum_{j=1}^{\infty}\sum_{k=j}^{\infty} \frac{jk}{2}(|\lambda_j(s) - \phi_j(s)||\lambda_k(s) + \phi_k(s)| + |\lambda_j(s) + \phi_j(s)||\lambda_k(s) - \phi_k(s)|)ds.$$

By changing the order of summation and some simplifications, one can obtain

$$u(\tau) \leq \int_0^{\tau} \left(\sum_{j=1}^{\infty}\sum_{k=j}^{\infty} k^3|\lambda_j(s) - \phi_j(s)||\lambda_k(s) + \phi_k(s)| + \sum_{j=1}^{\infty}\sum_{k=1}^{j} j^3|\lambda_j(s) + \phi_j(s)||\lambda_k(s) - \phi_k(s)| \right) ds$$

$$+ \int_0^{\tau} \left(\sum_{j=1}^{\infty}\sum_{k=j}^{\infty} \frac{k^2}{2}|\lambda_j(s) - \phi_j(s)||\lambda_k(s) + \phi_k(s)| + \sum_{j=1}^{\infty}\sum_{k=1}^{j} \frac{j^2}{2}|\lambda_j(s) + \phi_j(s)||\lambda_k(s) - \phi_k(s)| \right) ds.$$

Using the definition of $u(\tau)$ and Lemma 2.1, it is easy to observe that

$$u(\tau) \leq (4\beta + 2\alpha) \int_0^{\tau} u(s)ds.$$

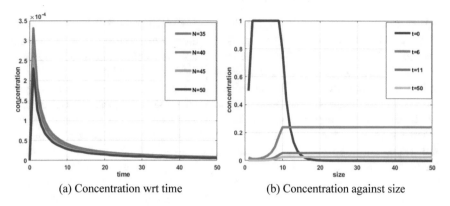

(a) Concentration wrt time　　　　　　　　(b) Concentration against size

Fig. 1 Concentration $\phi_j(\tau)$

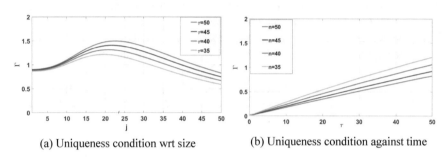

(a) Uniqueness condition wrt size　　　　　(b) Uniqueness condition against time

Fig. 2 Validation for $(H2)$

Hence, an application of Gronwall's lemma validates $u(\tau) \equiv 0$ when $0 \leq \tau \leq T$ and so $\phi_j(\tau) = \lambda_j(\tau) \ \forall j$ and $0 \leq \tau \leq T$. Since, T is arbitrary, this completes the proof of Theorem 3.1.

4　Numerical Simulations

The section illustrates the numerical solution of the discrete population balance Eq. (1) for the kernel $C_{j,k} = jk$. We have used fourth order-Runge–Kutta method to compute the numerical results. The numerical solution aids in plotting the second and third moments. Finally, these plots ensures the validation of $(H2)$ and confirm the existence of a unique solution. The hypothesis $(H1)$ is validated by computing the value of the second moment at various time intervals.

The following graphs are plotted by using ode45 in MATLAB. The other important parameters are the initial condition $\phi_j(0) = \frac{1}{2^j}$, $n = 50$ and $\tau \in [0, 50]$.

Figure 1 presents the numerical plot for the concentration $\phi_j(\tau)$ at different values of $\tau \in [0, 50]$ (Fig. 1a) and the normalized concentration for different sizes of particles is depicted through the Fig. 1b. Figure 1b clearly establishes the existence of a steady state, which can be the focus of a possible future work. Further, the uniqueness condition (H2) is validated for the numerical solution with the help of the Fig. 2. For the sake of convenience, the expression $\frac{7\alpha+5\beta}{12\alpha(\beta+1)}$ is denoted as Γ. In Fig. 2a, the values of Γ are plotted at different values of j for $\tau = 35, 40, 45, 50$. The figure depicts that the value of Γ increases first upto a maximum value and then decreases until the value becomes less than 1. This would mean that the solution is unique when $j \in [1, 8)$ and $j \in [43, 50]$ at $\tau = 50$. The plot of the linear combination of the second and third moments at these four values of τ seems to follow the same behaviour. The value of Γ is shown to be maximum at $\tau = 50$ and minimum at $\tau = 35$ for any fixed j. The plot of Γ, when drawn for $n = 35, 40, 45, 50$ indicates that since $45, 50 \in [43, 50]$, this implies that $\Gamma < 1$ and hence solution is unique for $n = 45, 50$ for every $\tau \in [0, 50]$. It also illustrates that as the value of n increases, the solution becomes unique for every τ. Figure 2 ensures that Γ is an increasing function wrt time but non-monotonic wrt size.

Now, that we have the validation for the hypothesis $(H2)$, the following table establishes that the hypothesis $(H1)$ also holds true for the example considered.

τ	0	10	20	30	40	50
$\lVert\phi\rVert_1$	1510	60.86	30.87	20.68	15.55	12.71

5 Concluding Remarks

The uniqueness of the solution required a condition on the finiteness of the second and third moments and then the relation between these finite quantities. The uniqueness condition helped us locate the values of j for which the solution is unique for every τ. Now, that the solution of the Safronov–Dubovski equation is unique for $C_{j,k} \leq jk$ $\forall\ j, k \in \mathbb{N}$, it would be interesting to study, in future, the analytical solution and steady-state solution for such model.

References

1. Dubovski PB (1999) A 'triangle' of interconnected coagulation models. J Phys A: Math. Gen 32(5):781
2. Dubovski PB (1999) Structural stability of disperse systems and finite nature of a coagulation front. J. Exp. Theor Phys 89(2):384–390
3. Bagland V (2005) Convergence of a discrete Oort-Hulst-Safronov equation. Math. Methods Appl. Sci. 28(13):1613–1632
4. Davidson J (2014) Existence and uniqueness theorem for the safronov-dubovski coagulation equation. Zeitschrift für Angewandte Mathematik und Physik 65(4):757–766
5. Davidson J (2016) Mathematical theory of condensing coagulation. PhD thesis, Stevens Institute of Technology

6. Oort JH, Van de Hulst HC et al (1946) Gas and smoke in interstellar space. Bull Astron Inst Netherlands 10:187
7. Safronov VS (1972) Evolution of the protoplanetary cloud and formation of the earth and the planets. Israel Program for Scientific Translations Jerusalem
8. Piotrowski SI (1953) The collisions of asteroids. AcA 6:115–138
9. Brilliantov N, Krapivsky PL, Bodrova A, Spahn F, Hayakawa H, Stadnichuk V, Schmidt J (2015) Size distribution of particles in saturn's rings from aggregation and fragmentation. Proc Nat Acad Sci 112(31):9536–9541
10. Wattis Jonathan AD (2012) A coagulation-disintegration model of Oort-Hulst Cluster-Formation. J Phys A: Math Theor 45(42):425001
11. Gross JE, Carlos WG, Cruz CSD, Harber P, Jamil S (2018) Sand and dust storms: acute exposure and threats to respiratory health. Am J Resp Crit Care Med 198(7):P13–P14
12. Gao W, Zhang X, Zhang D, Peng Q, Zhang Q, Dobashi R (2017) Flame propagation behaviours in nano-metal dust explosions. Powder Technol 321:154–162
13. Bradley RS, England J (1978) Volcanic dust influence on glacier mass balance at high latitudes. Nature 271(5647):736–738

Hybrid Modeling of COVID-19 Spatial Propagation over an Island Country

Jayrold P. Arcede, Rachel C. Basañez, and Youcef Mammeri

Abstract We propose a coupled metapopulation reaction-diffusion model to explain the propagation of the COVID-19 over an island country such as the Philippines. In the main islands, only susceptible, exposed, and asymptomatic individuals are traveling. The model takes into account the mean daily movement and the transfers and is parametrized using data on the confirmed cases and deaths from the Philippines. To proceed, we set up the system of partial and ordinary differential equations whereby the basic reproduction number, $\mathcal{R}_0$, is obtained. Next, we simulate the spatial spread of COVID-19 during the first 140 days of infection using a combination of level-set and finite difference methods. Afterward, scenarios of lockdown and unlockdown were studied. Our results displayed a remarkably close similarity to what happened in the Philippines during its first 140 days.

1 Introduction

A year has passed since January 30, 2020, where the Philippines' first index case of Corona Virus Disease 2019 (COVID-19) was recorded. However, the country continues to grapple with its impact on health, economic, social, and mental health [1, 12]. As of June 15, SARS-COV2, the virus that causes COVID-19, ravaged the whole world with over four (4) million deaths while infecting at least 176 million. In the Philippines, over a million infections had been confirmed, in which nearly 23 thousand had died while over one (1) million have recovered [2].

J. P. Arcede · R. C. Basañez
Department of Mathematics, Caraga State University, Butuan City, Philippines
e-mail: jparcede@carsu.edu.ph

R. C. Basañez
e-mail: rbapdo@carsu.edu.ph

Y. Mammeri (✉)
Laboratoire Amiénois de Mathématique Fondamentale et Appliquée, CNRS UMR 7352,
Université de Picardie Jules Verne, Amiens, France
e-mail: youcef.mammeri@u-picardie.fr

© The Author(s), under exclusive license to Springer Nature Singapore Pte Ltd. 2022   75
R. Srinivas et al. (eds.), *Advances in Computational Modeling and Simulation*,
Lecture Notes in Mechanical Engineering,
https://doi.org/10.1007/978-981-16-7857-8_7

In the early phase, the Philippines implemented the most extended lockdown. Yet, last April 2021, the country experienced a second wave of infections where cases reached as high as 15 thousand a day. The worst is that the infections crawled their way to the provinces where it now outweighs the number of cases from the National Capital Region (NCR), shifting epidemic hotspots to Visayas and Mindanao. Consequently, the government implemented a series of lockdowns with an average of two (2) weeks over the archipelago to flatten the curve. The sad reality, however, lockdowns only offer a short-term solution. Vaccination, regarded as the key to beat COVID-19, has been too slow for the Philippines (1.6% vaccinated individuals only as of June 14, 2021). Hence, to buy time, lockdowns were implemented by the government.

Mathematical modeling has been at the forefront in all decision-making related to COVID-19 [19], and there have been many papers related to COVID-19, which uses different mathematical tools to track down COVID-19 spread. We searched existing literature along this line and found that most proposed models are either discrete or continuous SIR type while a few only consider spatial spread. To mention some, Gardner [9] evaluated the expected number of cases in mainland China at the end of January 2020 using an SEIR-type metapopulation network. Wu et al. [22] simulated the COVID-19 thanks to an SEIR metapopulation model. Daily movements are considered by Danon et al. [7] and Giuliani et al. [10] utilized a statistical model to deal with the spread of COVID-19 in Italy. Recently, Mammeri [15] successfully applied a spatial propagation model in the case of France. We wonder if it can be adapted in an archipelago like the Philippines.

The rest of the paper is organized as follows. Section 2 outlines our model. Here, the qualitative analysis is done, and the computation of reproductive number $\mathcal{R}_0$ is provided. Section 3 deals with the numerical simulations. Section 4 concludes the paper.

2 Mathematical Model and Its Study

2.1 Description of the Model

In our study, five (5) components of the epidemic flow are examined (Fig. 1), i.e., the densities of Susceptible individual (S), Exposed individual (E), symptomatic Infected individual (I_s), asymptomatic Infected individual (I_a), and Removed individual (R). We assumed that susceptible could be infected by exposed and by infected individuals [3]. Further, we suppose that only susceptible, exposed, and asymptotic individuals are traveling. Following the literature concerning epidemic modeling [4], the dynamics is described in each island by a system of three partial differential equations (PDE) and three (3) ordinary differential equations (ODE) as follows, for $\mathbf{x} = (x, y) \in \Omega = \bigcup_{1 \leq i \leq l} \Omega_i \subset \mathbb{R}^2$, $t > 0$, and $1 \leq i \leq l$

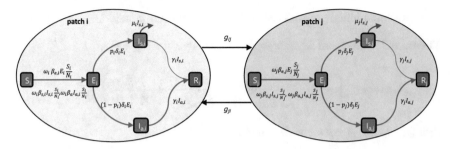

Fig. 1 Compartmental representation of the hybrid SEI_aI_sR−model. Blue arrows represent the infection flow. Purple arrows denote the death. Green compartments indicate moving individuals

$$
\begin{cases}
\partial_t S_i - d_i \Delta S_i = -\omega_i \left(\beta_{e,i} E_i + \beta_{s,i} I_{s,i} + \beta_{a,i} I_{a,i} \right) \dfrac{S_i}{N_i} + \sum_{j=1}^{l} g_{ji} S_j - g_{ij} S_i \\[2ex]
\partial_t E_i - d_i \Delta E_i = \omega_i \left(\beta_{e,i} E_i + \beta_{s,i} I_{s,i} + \beta_{a,i} I_{a,i} \right) \dfrac{S_i}{N_i} - \delta E_i + \sum_{j=1}^{l} g_{ji} E_j - g_{ij} E_i \\[2ex]
\partial_t I_{a,i} - \Delta I_{a,i} = (1 - p_i) \delta_i E_i - \gamma_i I_a + \sum_{j=1}^{l} g_{ji} I_{a,j} - g_{ij} I_{a,i} \\[2ex]
I'_{s,i} = p_i \delta_i E_i - (\gamma_i + \mu_i) I_{s,i} \\[1ex]
R'_i = \gamma_i (I_{a,i} + I_{s,i}).
\end{cases}
\tag{1}
$$

Homogeneous Neumann boundary conditions are imposed on islands, connectivity matrix $G = (g_{ij})$ translates travels between islands. The death is $D'(t) = \sum_i \mu_i I_{s,i}$ and the total living population is $N = S + E + I_a + I_s + R := \sum_i S_i + E_i + I_{a,i} + I_{s,i} + R_i$. The parameters are summarized in Table 1.

2.2 Initial Boundary Value Problem and Basic Reproduction Number

We suppose that all parameters are positive and bounded. Following [8, 15, 16], it can be proved that the model is globally well-posed.

Theorem 1 *For all $1 \leq i \leq l$. Let $0 \leq S_{0,i}, E_{0,i}, I_{a,i,0}, I_{s,i,0}, R_{0,i} \leq N_{0,i}$ be the initial datum with $\sum_i N_{0,i} \leq N_0$. Then there exists a unique global in time weak solution $(S_i, E_i, I_{a,i}, I_{s,i}, R_i)$ in $L^\infty(\mathbb{R}_+, L^\infty(\Omega))^{5l}$, of the initial boundary value problem. Moreover, the solution is nonnegative and $\sum_i S_i + E_i + I_{a,i} + I_{s,i} + R_i \leq N_0$.*

Moreover, if $d_i = d$ and $g_{ij} = \frac{g}{k_j}$, and $\omega_0 \beta_e \leq \delta$, then the solution converges almost everywhere to the Disease-Free Equilibrium $(S^*, 0, 0, 0, R^*)$ with $S^* + R^* = N^*$.

Exponential initial growth is ensured by the following condition.

Theorem 2 *If the local basic reproduction number, for $1 \leq i \leq l$,*

$$\mathcal{R}_{0,i} := \omega_{0,i} \left(\frac{\beta_{e,i}}{\delta_i + \sum_{j=1}^{l} g_{ij}} + \frac{(1-p_i)\beta_{a,i}}{\gamma_i + \sum_{j=1}^{l} g_{ij}} + \frac{p_i \beta_{s,i}}{\gamma_i + \mu_i} \right) \frac{S_{0,i}}{N_{0,i}} > 1,$$

then $(E_i, I_{a,i}, I_{s,i})$ exponentially grows.

The transmission rate due to exposed individuals during the average latency period $1/\delta_i$, accounting for all emigration g_{ij} of exposed individuals, is given by the term $\frac{\beta_{e,i}}{\delta_i + \sum_j g_{ij}}$. The dissemination due to asymptomatic individuals during the average infection period $1/\gamma_i$, accounting for all emigration g_{ij} of asymptomatic individuals, is given by $\frac{(1-p_i)\beta_{a,i}}{\gamma_i + \sum_j g_{ij}}$. The last term concerns those symptomatic individuals who are not traveling.

Proof A linearization around $(S_{0,i}, E_{0,i}, I_{a,i,0}, I_{s,i,0}, 0, 0)$ is written as

$$\begin{pmatrix} E_i \\ I_{a,i} \\ I_{s,i} \end{pmatrix}' = \begin{pmatrix} d_i \Delta + \omega_i \beta_{e,i} \frac{S_{0,i}}{N_{0,i}} - \delta_i - \sum_j g_{ij} & \omega_i \beta_{a,i} \frac{S_{0,i}}{N_{0,i}} & \omega_i \beta_{s,i} \frac{S_{0,i}}{N_{0,i}} \\ (1-p_i)\delta_i & d_i \Delta - \gamma_i - \sum_j g_{ij} & 0 \\ p_i \delta & 0 & -(\gamma_i + \mu_i) \end{pmatrix} \begin{pmatrix} E_i \\ I_{a,i} \\ I_{s,i} \end{pmatrix} + \begin{pmatrix} \sum_j g_{ji} E_j \\ \sum_j g_{ji} I_j \\ 0 \end{pmatrix}.$$

Let $(v_k)_{k \geq 1}$ be an orthonormal basis of eigenfunctions of the Laplace operator with the homogeneous Neumann boundary condition, i.e., $-\Delta v_k = k^2 v_k$. Writing $E_i, I_{a,i}, I_{s,i}$ as a linear combination of $(v_k)_{k \geq 1}$, it is therefore enough to study the characteristic polynomial reading as $P_{k,i}(x) = x^3 + a_2 x^2 + a_1 x + a_0$, with $a_0 = (\gamma_i + \mu_i)(d_i k^2 + \gamma_i + \sum_{j=1}^{l} g_{ij})(d_i k^2 + \delta_i + \sum_{j=1}^{l} g_{ij})(1 - \mathcal{R}_{k,i})$ and

$$\mathcal{R}_{k,i} := \omega_{0,i} \left(\frac{\beta_{e,i}}{d_{0,i} k^2 + \delta_i + \sum_{j=1}^{l} g_{ij}} + \frac{(1-p_i)\beta_{a,i}}{d_{0,i} k^2 + \gamma_i + \sum_{j=1}^{l} g_{ij}} + \frac{p_i \beta_{s,i}}{\gamma_i + \mu_i} \right) \frac{S_{0,i}}{N_{0,i}}.$$

At least one (1) positive eigenvalue exists if $\mathcal{R}_{0,i} > 1$, which leads to the initial exponential growth rate of solutions. □

Remark 1 Infection is depicted by the individuals in $E_i, I_{a,i}$ and $I_{s,i}$, new infections $(\mathcal{F}_i)$ and transitions between compartments $(\mathcal{V}_i)$ are

$$\mathcal{F}_i = \begin{pmatrix} \omega_i (\beta_{e,i} E_i + \beta_{s,i} I_{s,i} + \beta_{a,i} I_{a,i}) \frac{S_i}{N_i} \\ 0 \\ 0 \end{pmatrix}, \quad \mathcal{V}_i = \begin{pmatrix} \delta_i E_i - \sum_{j=1}^{l} g_{ji} E_j + g_{ij} E_i \\ \gamma_i I_{a,i} - (1-p_i)\delta_i E_i - \sum_{j=1}^{l} g_{ji} I_j + g_{ij} I_i \\ (\gamma_i + \mu_i) I_{s,i} - p_i \delta_i E_i \end{pmatrix},$$

then the local basic reproduction number $\mathcal{R}_{0,i} := \rho(F_i V_i^{-1})$ [21].

3 Numerical Results

3.1 Numerical Discretization

To simply discretize an island country, the level-set [18] is defined as the function ϕ such that the territory $\Omega := \{\mathbf{x} \in \mathbb{R}^2; \ \phi(\mathbf{x}) < 0\}$, and its boundary is the zero level of ϕ. The exterior normal is $\mathbf{n} = \frac{\nabla\phi}{||\nabla\phi||}$. The computation domain consists of a Cartesian grid which is given by the image pixels (Fig. 2). Then the map of population density allows the building of initial population density, $N_0(\mathbf{x})$ (Fig. 2). Finally, Runge-Kutta fourth order discretizes the time derivative and the second-order central finite difference solves the space derivative.

3.2 Parameters Calibration

Latency period and infection period have been evaluated as five (5) days and seven (7) days, respectively [13], and thus $\delta = 1/5, \gamma = 1/7$. Local lockdowns and new variants are computed by updating the average number of contacts [14, 15]

$$
\omega_i(t) = \begin{cases} \omega_{0,i} \text{ if } t \leq t_{lockstart,i} \\ \omega_{0,i}e^{-\rho_{l,i}(t-t_{lockstart,i})} \text{ if } t_{lockstart,i} \leq t \leq t_{lockend,i} \\ \dfrac{\eta_i\omega_{0,i}}{1 + (\eta_i e^{\rho_{l,i}(t_{lockend,i}-t_{lockstart,i})} - 1)e^{-\rho_{u,i}(t-t_{lockend,i})}} \text{ if } t \geq t_{lockend,i}, \end{cases}
$$

(2)

and by updating the diffusion coefficient

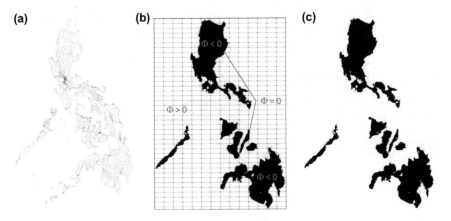

Fig. 2 **a** Map of population density of the Philippines. **b** Level-set and computation domain consisting in the Cartesian grid. **c** Five (5) main islands of the Philippines and its airports as red dots

$$
d_i(t) =
\begin{cases}
d_{0,i} & \text{if } t \leq t_{lockstart,i} \\
d_{0,i} e^{-\rho_{l,i}(t-t_{lockstart,i})} & \text{if } t_{lockstart,i} \leq t \leq t_{lockend,i} \\
\dfrac{d_{0,i}}{1 + (e^{\rho_{l,i}(t_{lockend,i}-t_{lockstart,i})} - 1)e^{-\rho_{u,i}(t-t_{lockend,i})}} & \text{if } t \geq t_{lockend,i}.
\end{cases}
\quad (3)
$$

Here, $lockstart, i$ denotes for start of lockdown and $lockend, i$ for end of lockdown in island i. The parameter η_i provides the infection multiplier rate according to the compliance with distancing rules, or introduction of new variants in patch i.

Since data are not fully available in each island, parameters $\theta = (\rho_l, \rho_u, \eta, \omega_{i,0}\beta_e, \omega_{i,0}\beta_s, \omega_{i,0}\beta_a, p, \mu)$ are calibrated for the whole country using approximate Bayesian computation of the nonlinear least square cost function.

The COVID-19 dataset is openly provided by Philippine government. The dataset can be downloaded from https://data.gov.ph. The first confirmed infection was on January 30, 2020. We remind that lockdown in the Philippines started on March 15, i.e., 45 days after the first infection, then extended until May 31 and finally reimposed on August 4. The daily commuting is set to 15 km and the value of d_0 is fixed equal to $\frac{15^2}{16}$ [20]. Table 1 shows the mean estimated parameters after 2000 runs.

Table 1 List of parameters and its fitted value

J	Relative cost function	0.00218
i	Patch number	$1-5$
d_i	Diffusion coefficient	$\frac{15^2}{16}$
ω_i	Contact rate	Not identifiable
β_i	Probability transmission	Not identifiable
$\omega_{i,0}\beta_{e,i}$	Transmission rate from S_i to E_i from contact with E_i	0.0748
$\omega_{i,0}\beta_{s,i}$	Transmission rate from S_i to E_i from contact with $I_{s,i}$	0.2879
$\omega_{i,0}\beta_{a,i}$	Transmission rate from S_i to E_i from contact with $I_{a,i}$	0.3381
η_i	Infection multiplier in patch i	0.5179
$\rho_{l,i}$	Exponential decay for lockdown	0.0094
$\rho_{u,i}$	Exponential decay for unlockdown	0.8013
δ_i	Latency rate	1/5
p_i	Probability of being symptomatic	0.4388
$1-p_i$	Probability of being asymptomatic	0.5612
γ_i	Recovery rate	1/7
μ_i	Death rate	0.00016
$\mathcal{R}_0$	Basic reproduction number	2.5435

3.3 Spatial Spread of COVID-19

We consider the five (5) main islands and five (5) main airports in the Philippines. The connectivity coefficients g_{ij} are given as the number of passengers traveling between i and j, divided by the number of passengers in i. Because the data are not openly accessible, we assume that the very first infected individuals were located at Manila Airport on day one (1), and moved to each of the six (6) chosen airports on day two (2). This is the only assumed travel during the simulation.

The day before lockdown is depicted in Fig. 3-A. The disease is mostly located in the NCR, Western Visayas, and a big part of Mindanao encompassing Davao Region, Socksargen, Northern Mindanao, and Caraga. After 92 days, (Fig. 3-B1), the disease persists in these regions. It almost vanishes in small islands. Given that lockdowns continue, the density of symptomatic infected individuals reduces until day 140 but completely extinguished in all of Visayas and Mindanao. At the same time, it shifts to the central part of Luzon, i.e., Cordillera Region, Cagayan Valley, and Ilocos Region (Fig. 3-B2).

Figure 3-C depicts the situation without lockdown. After 92 days, the disease is in the same regions, but with many infected individuals. As a result, the propagation

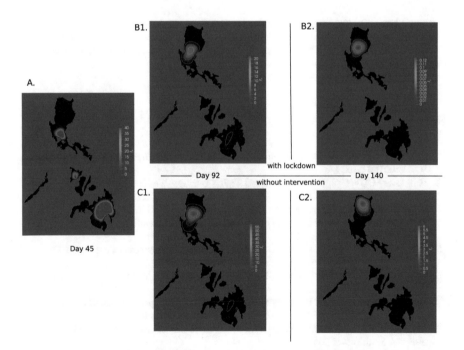

Fig. 3 **A** Spatial distribution of symptomatic infected density on day 45. **B** Spatial distribution of symptomatic infected density on day 92 (B1) and on day 140 (B2) when lockdown started on day 45. **C** Spatial distribution of symptomatic infected density on day 92 (C1) and on day 140 (C2) without lockdown

moves to the north. While the disease vanishes in the southern part of the country, after 140 days the north is peaking.

4 Conclusion

We have presented a spatiotemporal model for COVID-19 spread in the Philippines. Remarkably, the simulation pattern captured the actual spread of COVID-19 dynamics in the country. That is, with or without lockdown and discounting delay, the propagation focused around NCR where dense cities like Quezon City and Manila served as the epicenters [17]. After that, the disease spreads simultaneously to the other following big islands of Visayas (Cebu City as a hotspot) [6] and Mindanao (Davao City as a hotspot) [5] then it goes back to Luzon again to the north with Mountain Province experiencing some cluster [11].

The proposed model is capable of generating maps describing the spatiotemporal propagation of the epidemic for island country. Provided that data on movement between patches is available, it would be possible to provide a global as well as a local policy support with maps embedded with effective reproduction numbers.

Acknowledgements This work was supported by DOST-PCHRD under the project Understanding COVID-19 Pandemic Situation in Caraga Region through Epidemiological Models and Resiliency Studies (UnCOVER) 2020. YM is funded by the Agence National de la Recherche and Région Hauts-de-France, projet Space-covid ANR Résilience.

References

1. COVID-19: an ongoing public health crisis in the Philippines. Lancet Reg Health—Western Pac 9:100160 (2021)
2. Philippines COVID: 1,315,639 Cases and 22,788 Deaths—Worldometer (2021)
3. Arcede JP, Caga-anan RL, Mentuda CQ, Mammeri Y (2020) Accounting for symptomatic and asymptomatic in a seir-type model of covid-19. Math Mod Nat Pheno
4. Brauer F, Castillo-Chavez C (2012) Mathematical models in population biology and epidemiology. Springer, New York, NY
5. Crisostomo S, Romero A, Mendez C (2020) Nine mindanao provinces emerging as COVID-19 Hotspots; Gov't seeks ways to open up the economy
6. Dancel R (2020) Covid-19 'hot spot' Cebu City to remain on lockdown; manila restrictions eased further, July 2020
7. Danon L, Brooks-Pollock E, Bailey M, Keeling MJ (2020) A spatial model of covid-19 transmission in england and wales: early spread and peak timing. MedRxiv
8. Fitzgibbon WE, Langlais M, Morgan JJ (2007) A mathematical model for indirectly transmitted diseases. Math Biosci 206:233–248
9. Gardner L (2020) Modeling the spreading risk of 2019-ncov. Technical report, Center for Systems Science and Engineering, Johns Hopkins University
10. Giuliani D, Dickson MM, Espa G, Santi F (2020) Modelling and predicting the spatio-temporal spread of coronavirus disease 2019 (covid-19) in Italy

11. Gonzales C (2021) DOH reports 8 more cases of UK Covid-19 variant; total in PH now 25, Feb 2021
12. Kahambing JGS (2021) Psychosocial wellbeing and Stress Coping Strategies during COVID-19 of Social Workers in Southern Leyte, Philippines. Asian J Psychiatr 102733 (2021)
13. Lauer SA, Grantz KH, Bi Q, Jones FK, Zheng Q, Meredith HR, Azman AS, Reich NG, Lessler J (2019) The incubation period of coronavirus disease, (covid-19) from publicly reported confirmed cases: estimation and application. Ann Int Med 03:2020
14. Liu Z, Magal P, Seydi O, Webb G (2020) Predicting the cumulative number of cases for the covid-19 epidemic in china from early data. Math Bios Eng 17:3040
15. Mammeri Y (2020) A reaction-diffusion system to better comprehend the unlockdown: application of seir-type model with diffusion to the spatial spread of covid-19 in France. Comput Math Biophys 8(1):102–113
16. McCormack RK, Allen LJS (2007) Multi-patch deterministic and stochastic models for wildlife diseases. J Bio Dyn 63–85
17. Mercado NA (2020) Makati, Quezon City remain as virus hotspots; localized lockdowns pushed, Dec 2020
18. Osher S, Fedkiw R (2002) Level set methods and dynamic implicit surfaces. Springer, New York
19. Panovska-Griffiths J (2020) Can mathematical modelling solve the current Covid-19 crisis? BMC Publ Health 20(1):551
20. Shigesada N, Kawasaki K (1997) Biological invasions: theory and practice. Oxford University Press
21. van den Driessche P, Watmough J (2000) Reproduction numbers and sub-threshold endemic equilibria for compartmental models of disease transmission. Math Biosci 180(1–2):29–48
22. Wu JT, Leung K, Leung GM (2020) Nowcasting and forecasting the potential domestic and international spread of the 2019-NCOV outbreak originating in Wuhan, China: a modelling study. The Lancet 395(10225):689–697

Identification of Potential Inhibitors Against SARS-CoV-2 3CL^{pro}, PL^{pro}, and RdRP Proteins: An *In-Silico* Approach

Manju Nidagodu Jayakumar, Jisha Pillai U., Moksha Mehta, Karanveer Singh, Eldhose Iype, and Mainak Dutta

Abstract The COVID-19 pandemic is still evolving and is caused by SARS-CoV-2. Me-too drugs are chemically, structurally, or functionally similar to preexisting drugs. These drugs may be chemically related to the prototype, and may have an identical mechanism of action with enhanced target specificity and reduced risks of adverse reactions. In the present study, we have performed a chemical similarity analysis to identify Me-too (similar) ligands relative to previously reported potential drug hits (reference compounds) against the major viral proteins, including 3CL^{pro}, PL^{pro}, and RdRP. The binding efficiency of the similar molecules was then evaluated by using molecular docking and molecular dynamics simulation approaches to assess their binding effectiveness relative to the reference compounds. Our results indicate that several molecules show better interaction compared to the reference molecules. These molecules may be potential drugs inhibiting different stages of the SARS-CoV-2 viral lifecycle.

Keywords SARS-CoV-2 · COVID-19 · 3CL^{pro} · PL^{pro} · RdRP · Drug discovery

---

Electronic supplementary material The online version of this chapter (https://doi.org/10.1007/978-981-16-7857-8_8) contains supplementary material, which is available to authorized users.

M. N. Jayakumar · J. Pillai U. · M. Mehta · K. Singh · M. Dutta (✉)
Department of Biotechnology, Dubai Campus, BITS Pilani, Academic City, Dubai, United Arab Emirates
e-mail: mainak@dubai.bits-pilani.ac.in

M. N. Jayakumar
Sharjah Institute of Medical Research, University of Sharjah, Sharjah, United Arab Emirates

E. Iype
Department of Chemical Engineering, Dubai Campus, BITS Pilani, Academic City, Dubai, United Arab Emirates

1 Introduction

The global outbreak of the current coronavirus disease 2019 (COVID-19) is caused by severe acute respiratory syndrome coronavirus 2 (SARS-CoV-2). The virus belongs to the *Coronaviridae* family of *Betacoronavirus* genus (Coronaviridae Study Group of the International Committee on Taxonomy of Viruses, 2020). In the past two decades, two other coronaviruses of the same genus, SARS-CoV and MERS-CoV, had caused two major worldwide outbreaks [1]. The World Health Organization (WHO) declared COVID-19 a pandemic in March 2020 [2]. The disease had resulted in >216 million cumulative cases globally causing more than 4.49 million deaths as of 30th August, 2021 [3]. Though efforts are being made for large-scale vaccination drives in majority of the countries, additional threats from mutated variants of the virus [4], incidents of infection in vaccinated individuals, possibilities of reinfection in recovered patients [4, 5], and short duration of seropositivity for neutralizing antibodies [4] in vaccinated individuals raise alarm about the future course of the current COVID-19 pandemic.

The coronavirus genome encodes four structural proteins, spike (S), nucleocapsid (N), membrane (M) and envelope (E) proteins, and 16 non-structural proteins (NSPs). Spike is a transmembrane homotrimeric glycoprotein that protrudes from the virus surface [6] and is responsible for host cell recognition and viral fusion. The N protein coats the positive single-stranded RNA genome of the virus forming a capsid. It protects the viral genome from the harsh environment of the host cell [7]. M protein is an integral membrane protein that determines the shape of the viral envelope and interacts with S and N proteins [8]. The E protein is the smallest structural protein that acts as a scaffold to facilitate viral assembly and release [9]. Once the virus is inside the host cell, the replicase gene present in the viral genomic RNA is translated. This gene encodes two polyproteins, pp1a and pp1ab, that are cleaved into individual NSPs. Two proteases, main protease, 3CL^{pro} (NSP5) and papain-like protease, PL^{pro} (NSP3), cleave the polyproteins to form other NSPs. Many of these NSPs, including the primary RNA-dependent RNA polymerase (RdRP; NSP12), NSP7–NSP8 primase complex, a helicase–triphosphatase (NSP13), an exoribonuclease (NSP14), an endonuclease (NSP15), and N7- and 2′O-methyltransferases (NSP10 and NSP16) then form a replicase–transcriptase complex (RTC) which is ultimately involved in RNA replication and transcription [10].

In-silico approaches for virtual screening of ligands have emerged as a powerful tool in drug discovery. Since the outbreak of COVID-19, several studies have used a molecular docking approach to screen for drugs against SARS-CoV-2, which either target the virus entry into the host cell or the viral replication and transcription machinery [11–18]. Since the early phase of the pandemic, several such drugs have been undergoing clinical trials worldwide for their efficacy against the disease [19]. However, a number of trials were discontinued due to limited evidence for viral efficacy [20–23]. Chemical similarity analysis is another *in-silico* approach for potential drug identification. The systematic exposition of chemical network analysis offers the possibility of faster designing and reduced complexity in drug discovery [24, 25].

This approach is capable of integrating with biological knowledge-based databases (Uniprot, GO) and high-throughput biology platforms (proteomic, genetic, etc.) for system-wide drug target validation [26]. It generates "Me-too" drugs that are chemically, structurally, or functionally similar to preexisting drugs. These drugs may be chemically related to the prototype, and may have an identical mechanism of action. More than 60% of medicines listed on the WHO's essential list are Me-too drugs. Advantages of Me-too drugs include enhanced target specificity, reduced risks of adverse reactions and drug–drug interactions, increased chance of benefit in some patients, and improved drug delivery [27].

In the present study, we have performed a chemical similarity analysis to identify Me-too (similar) ligands relative to the previously reported potential drug hits (reference compounds) against the major viral proteins, including 3CL^{pro}, PL^{pro}, and RdRP. The binding efficiency of the similar molecules was then evaluated by using the molecular docking approach to assess their binding effectiveness with relative to the reference compounds. Finally, the stability of a lead protein–ligand complex was validated using molecular dynamics simulation.

2 Material and Methods

An overview of the methodology of the current work is represented in Fig. 1 and explained in details below.

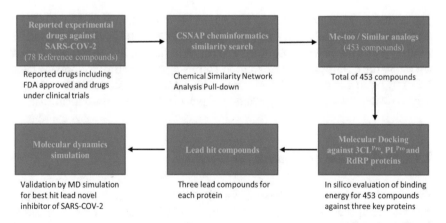

Fig. 1 Schematic flow chart of overall workflow of the *in-silico* Me-too drug repurposing pipeline

2.1 Chemical Similarity Analysis

Ligands, which are under clinical trial or have been previously reported, till August 2020, as potential drug targets against SARS-CoV-2, were selected as reference compounds. Networks of similar ligands relative to the reference compounds were generated using the web-based tool for Chemical Similarity Network Analysis Pull-down (CSNAP) analysis (https://services.mbi.ucla.edu/CSNAP/index.html) [24]. CSNAP is a network-based approach for automated compound target identification based on the current artificial intelligence and chemical similarity graphs [25, 26]. The FP2 parameter was chosen for both search and cluster fingerprints. This parameter has high specificity and is applicable to any ligand size. Structurally similar ligands were retrieved from the CHEMBL database version 20 using Tanimoto coefficient (Tc) and Z-scores. Tc is the most widely used parameter for chemical similarity analysis in cheminformatics studies. It is represented as the ratio between conserved features and the total number of features for each molecule. Its values range from 0 (no similarity) to 1 (total similarity). The statistical significance of this similarity coefficient is assessed using Z-score, which is based on the overall mean and standard deviation of score distribution of the hits [28]. In our current study, we have used a Tc and Z-score cutoff of 0.85 and 2.5, respectively, to identify the Me-too molecules.

2.2 Preparation of the Protein Targets for Docking

The crystal structures of 3CL^{pro} (PDB ID: 6Y2E; resolution: 1.75 Å) [29], PL^{pro} (PDB ID: 6WX4; resolution: 1.66 Å) [30], and RdRP (PDB ID: 6M71; resolution: 2.90 Å) [31] were obtained from the Protein Data Bank at the RCSB site (http://www.rcsb.org). The proteins were preprocessed using Autodock MGL Tools (Version 1.5.6) by removing co-crystallized water, inhibitors, and small molecules. Further preprocessing was performed by addition of polar hydrogens and Kollman charge.

2.3 Preparation of Ligands for Docking

Two-dimensional (2D)/three-dimensional (3D) SDF files of the reference ligands and their similar ligands were downloaded from the PubChem database. Web version of the Open Babel tool was used to convert the 2D structures to the 3D format and downloaded as.pdb file. The 3D structures were further manually verified using either PyMol (The PyMOL Molecular Graphics System, Version 2.0 Schrödinger, LLC), UCSF Chimera version 1.14 [32], or Avagadro [33]. Finally, polar hydrogens and Gasteiger charge were applied to all of the ligands using Autodock MGL Tools (Version 1.5.6) and saved as .pdbqt files.

2.4 Molecular Docking

Receptor grid boxes were generated by AutoGrid4 around the active site of proteins. Grid-box dimensions for each of the proteins are as follows: $100\text{ Å} \times 100\text{ Å} \times 100\text{ Å}$ and 0.2 Å spacing were used for 3CL^{pro}, PL^{pro}, and RdRP. 3CL^{pro} grid was centered at $X = -13.553$, $Y = -27.671$, and $Z = 0.825$ around amino acid residues GLN189, HIS163, GLU166, CYS145, and HIS41 [29]. The grid for PL^{pro} was centered at $X = 12.314$, $Y = -23.045$, and $Z = -40.204$ around the amino acid residues CYS111, GLY163, HIS272, GLN269, ASP286, TYR273, and ASP302 [30]. A grid center size of $X = 123.910$, $Y = 113.340$, $Z = 122.605$ around amino acid residues ASP618, ASP760, and SER759 was used for RdRP [34, 35]. The genetic algorithm (GA) search parameter was used for docking in combination with the grid-based energy evaluation method. A total number of 10 GA were used to run the docking program. The molecular docking results for protein–ligand interactions were evaluated using the PyMOL Molecular Graphics System and Discovery Studio (Dassault Systems BIOVIA).

2.5 Molecular Dynamics Simulations

Molecular dynamics simulations of RdRP protein with the Remdesivir similar compound-155 complex were performed using GROMACS package [36] with the CHARMM36 [37] force field starting from the docked state of the drug. A cubic box (with 10 Å additional space in all directions from the ends of the protein–ligand complex) filled with TIP3P [38] water molecules and Na^{+}/Cl^{-} atoms to neutralize charges were used. After energy minimization, the complex was allowed to equilibrate for 1 ns each under NVT and NPT ensemble with position restraints attached to ensure that the complex is stable. Then, a production run was performed for up to 70 ns. The MD trajectories have been saved every 10 ps for the entire simulation. The time step is 2 fs. During equilibration, position restraints were applied to both ligands and proteins so that there are no drastic changes in the structure during equilibration. The root mean square deviations (RMSD) of the protein and the ligand was calculated with respect to the starting structure of MD production run to understand the stability of the complex. In order to understand the binding energies, a molecular mechanics Poisson–Boltzmann surface area (MM–PBSA) analysis is performed over a selected number of frames taken between 50 and 70 ns simulations. The formation of hydrogen bonds between the protein and ligands was analyzed using an in-house code, where we identify the residues that are closer to the ligand by at the most 15 Å, followed by scanning for hydrogen bonds between polar atoms (N and O) with H atoms with a maximum interaction length of 2.5 Å.

3 Results and Discussion

With increasing number of cases and deaths, the current COVID-19 pandemic requires immediate therapeutic modalities. For the past several months, studies have identified potential drugs against SARS-CoV-2 proteins using virtual screening approach. Several of these molecules showed encouraging *in vitro* activity against the virus [12]; however, many of them are thought to have a poor selectivity index, indicating that higher tolerable levels of the drug might be required to achieve meaningful inhibition in vivo [39]. This is further supported by the fact that two such molecules, Lopinavir and Ritonavir, showed encouraging *in vitro* activity against coronaviruses in the earlier studies [39]. However, recent studies show limited efficacy in randomized controlled trials for SARS-CoV-2 [22]. Similarly, recent studies on hydroxychloroquine also showed limited efficacy [21]. Over the next few months when more clinical trials will reach the final stage, several potential drugs are expected to show limited efficacy against the disease. It is therefore imperative to identify new molecules with improved binding efficacy against the viral proteins. In the present study, we attempt to perform virtual screening to identify ligands having better binding affinity with the viral proteins compared to previously reported potential inhibitors.

3.1 Chemical Similarity Analysis

CSNAP is a powerful computational target identification method that utilizes chemical similarity networks for large-scale chemotype recognition and drug target profiling [26]. Me-too drugs were identified using CSNAP against previously reported potential COVID-19 drugs (reference compounds). We carefully selected these reference compounds from the peer-reviewed literature using *in-silico* approaches for virtual screening of ligands to novel SARS-CoV-2 [11, 13, 40–43]. Shortlisted compounds for our study included FDA-approved drugs or drugs under clinical trials. Our chemical similarity analysis of 78 reference compounds resulted in more than 453 similar Me-too molecules, with the Tanimoto coefficient (Tc) score and a Z-score cutoff of 0.85 and 2.5, respectively **(Supplementary File A)**. These compounds are structurally similar and might possess a similar mechanism of action to that of their reference drugs [44]. Molecular docking analysis was then performed with all of the ligands against their corresponding proteins to assess their binding efficacy.

3.2 Inhibitors of 3CL^{pro}

3CL^{pro} consists of 306 amino acids and cleaves at 11 different sites of the polyprotein to produce NSP4 to NSP16. It is essential for the lifecycle of virus and hence is considered as a potent drugable target [14]. During the preparation of this manuscript, evidences show that more than 18 experimental drugs were repurposed for targeting 3CL^{pro} and some are under clinical trials. These include acefluranol, benaxibine, tobramycin, sulfamoxole, sulfaethidole, sulfametrole, pentisomicin, glybuthiazol, tanzisertib, ramifenazone, etersalate, alpelisib, bisantrene, menatetrenone, bucindolol, mannitol, curcumin, and RO-24-7429 [11, 14–17, 45–48] (**Supplementary Table S1 A**). Using CSNAP analysis with these drugs as the main reference compounds we identified a total of 70 similar compounds against 3CL^{pro}. A complete list of the compounds can be found in **Supplementary File A**. The binding energy of these similar ligands generated from molecular docking analysis shows that the Me-too compounds either bind similarly or more effectively to the active site of the enzyme compared to the respective reference molecules (**Supplementary File A, Supplementary Table S1 B**). It is worthwhile to mention that compound-302 (−9.96 kcal/mol), compound-361 (−9.87 kcal/mol), and compound-368 (−10.7 kcal/mol) showed improved binding energy compared to their corresponding reference compounds acefluranol (−8.67 kcal/mol), alpelisib (−8.51 kcal/mol), and bisantrene (−9.4 kcal/mol), respectively (Fig. 2).

Acefluranol is an estrogen antagonist and is used as an estrogen blocker. It is used as a combination therapy for the treatment of estrogen-sensitive diseases [49]. CSNAP cheminformatics analysis of acefluranol resulted in a lead similar molecule, compound-302 (Scheme 1, Fig. 2a). Molecular docking analysis revealed that this similar molecule showed a better docking score compared to acefluranol (Fig. 2d).

Upon analyzing the 3D and 2D interaction maps between the ligand and 3CL^{pro}complex, it was found that compound-302 forms six hydrogen bonds with the amino acid residues, GLN192, THR190, SER144, ASN142, and CYS145 (Fig. 3a, b, Table 1). In contrast, acefluranol interacts with the protein with three hydrogen bonds (**Supplementary Figure S1A, S1B, Supplementary Table S2**). Further, an additional methyl group in compound-302 helps it to interact with 3CL^{pro} using an extra pi-alkyl interaction with MET49 (Fig. 3b, **Supplementary Figure S1B**) leading to its improved binding affinity compared to acefluranol.

Alpelisib is a FDA-approved drug for the treatment of breast cancer. It is a phosphatidylinositol 3-kinase (PI3K) inhibitor that specifically inhibits the PI3K/AKT kinase (or protein kinase B) signaling pathway [50–52]. CSNAP cheminformatics analysis of alpelisib resulted in seven similar compounds (Fig. 2b). Among them, compound-361 was considered as the lead molecule because it showed the highest docking score (Scheme 2, Fig. 2e). A higher score can be explained by the fact that this molecule interacts with GLY143 with a strong conventional hydrogen bond (1.78 Å), compared to alpelisib (1.84 Å). Additionally, the absence of a fluoride group in compound-361 allows it to orient in a way that it forms the alkyl interaction with

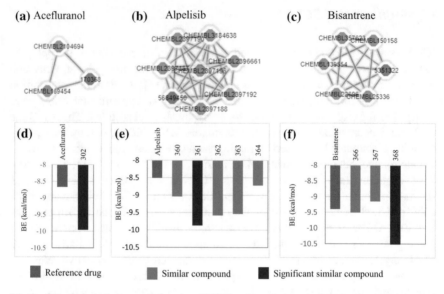

Fig. 2 Chemical similarity analysis using CSNAP to identify Me-too drugs against SARS-CoV-2 3CL^{Pro}. Similarity network map for reference repurposed drugs **a** acefluranol, **b** alpelisib, and **c** bisantrene. **d–f** Column plots representing molecular docking score of the Me-too drugs compared to the corresponding reference drugs. (*Note*: Ligands that could not be docked due to structural complexity are not included in the bar graph.)

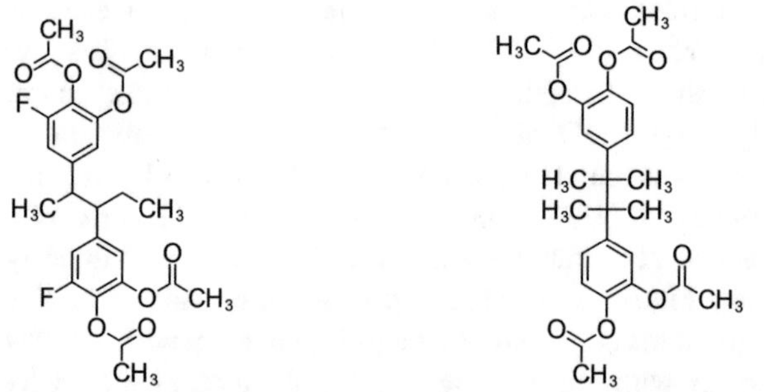

Acefluranol (Pubchem# 170368)   Compound-302 (Pubchem# 12610668)

Scheme 1 Chemical structure of acefluranol and compound-302

LEU27 (Fig. 3c, d) compared with alpelisib (**Supplementary Figure S1C, S1D**). Other interactions involved can be found in Table 1 and **Supplementary Table S2.**

Bisantrene is an antineoplastic drug that intercalates with and disrupts the configuration of DNA, thereby inhibiting DNA replication [53]. CSNAP cheminformatics analysis of bisantrene resulted in five similar compounds (Fig. 2c). Among them, compound-368 was considered as the lead molecule because it showed the highest docking score (Scheme 3, Fig. 2f). A higher score can be explained by the fact that this

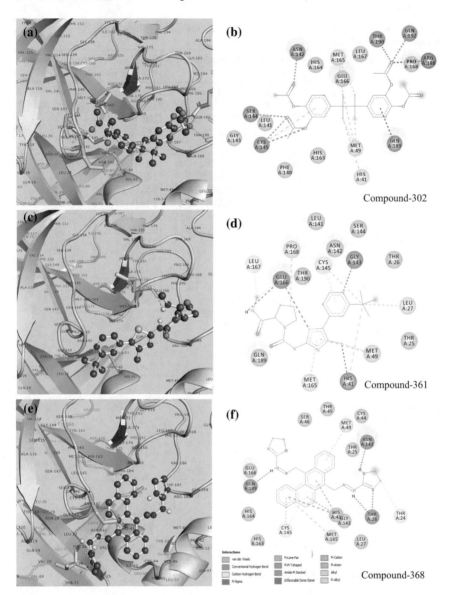

Fig. 3 Low-energy binding conformations of the SARS-CoV-2 3CL^Pro–ligand docked complex showing 3D interactions (**a, c, e**) created by Pymol, 2D interaction (**b, d, f**) of ligand with amino acid residues of 3CL^Pro created by discovery studio for compounds-302, 361, and 368, respectively

Table 1 List of interacting amino acid residues of SARS-CoV-2 3CL^{Pro} with selected novel lead compounds-302, 361, and 368

Compound	PubChem ID	Hydrogen bonds	Hydrophobic interaction	Other interactions	Binding energy (kcal/mol)
302	12610668	GLN192, THR190, CYS145, SER144, ASN142	GLN189, MET165, MET49, HIS41, CYS145	ARG188, LEU141, HIS163, HIS164	−9.96
361	71682652	GLU166 , GLY143	PRO168, MET165, MET49, HIS41, LEU27, CYS145	GLU166, LEU167, MET49, LEU141, ASN142, SER144, THR26, THR35, GLN189	−9.87
368	44365428	ASN142, THR26, GLN189	CYS145, HIS41, MET49, MET165	THR24, SER46, THR45, CYS44, LEU27, HIS163, HIS164	−10.7

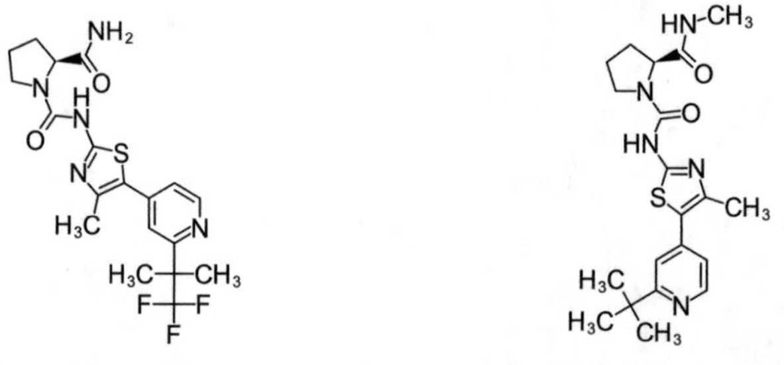

Alpelisib (PubChem# 56649450)   Compound-361 (PubChem# 71682652)

Scheme 2 Chemical structure of alpelisib and compound-361

molecule interacts with 3CL^{pro} amino acid residues, ASN142, THR26, and GLN189, via three hydrogen bonds (Fig. 3e, f, Table 1), whereas bisantrene is involved in two such interactions (**Supplementary Figure S1E, S1F, Supplementary Table S2**). Further, an extra benzyl group in compound-368 helps it to interact with the enzyme by additional hydrophobic and van der Waals interactions (Fig. 3f).

Additionally, all of these lead compounds were verified for binding specific to the active site by comparing the full ribbon structure of the receptor and ligand docked complex along with their reference drugs (Fig. 4).

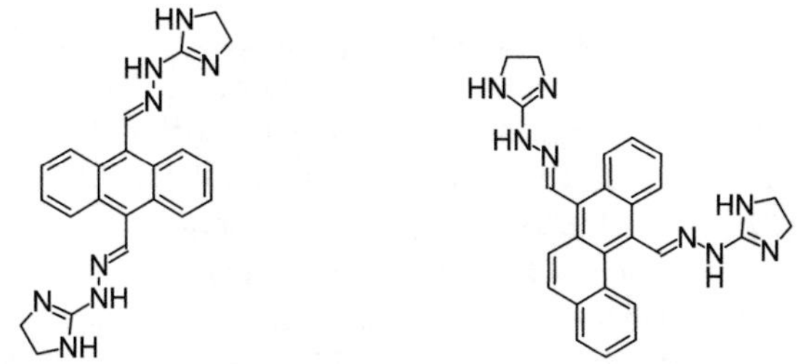

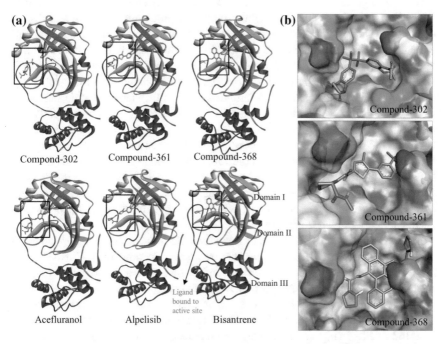

Bisantrene (PubChem# 5351322)   Compound-368 (PubChem# 44365428)

Scheme 3 Chemical structure of bisantrene and compound-368

Fig. 4 Docked protein complex of top hit compounds for SARS-Cov-2 3CL^{Pro} protein. **a** Ribbon structure of docked protein complex of top hit compounds-302, 361, and 368 with their reference drugs alpelisib, acefluranol, and bisantrene, respectively, showing ligands docked at the active pocket (highlighted in square box). **b** Surface protein image and active site binding of compounds-302, 361, and 368 surface colored by hydrogen bond type

3.3 Inhibitors of Papain-like Protease (PL^{pro})

PL^{pro} is a large multi-domain membrane-bound protein with 194 amino acids. It cleaves pp1a to generate three NSPs 1, 2, and 3, as products. It is considered as a drugable target, as it not only is involved in active replication of virus, but also negatively regulates antiviral innate immune response [54]. During the preparation of this manuscript, reports show that more than 36 experimental drugs, including FDA-approved drugs and drugs under clinical trials, were repurposed for targeting PL^{pro}. These include praziquantel (biltricide), cinacalcet, procainamide, terbinafine, meperidine, labetalol, tetryzoline, ticlopidine, ethoheptazine, levamisole, amitriptyiline, naphazoline, arformoterol, benzylpencillin, chloroquine, chlorothiazide, ribavirin, valganciclovir, aspartame, oxprenolol, acetophenazine, iopromide, riboflavin, reproterol, chloramphenicol, chlorophensin carbamate, levodropropizine, cefamandole, floxuridine, pemetrexed, glutathione, hesperetin, ademethionine, masoprocol, isotretinoin, silybin, and nicardipine [17, 55–61] (**Supplementary Table S3 A**). Using these drugs as reference compounds, a similar compound network was generated by CSNAP. A total of 159 compounds were shortlisted from CSNAP analysis for molecular docking studies (**Supplementary File A**). Molecular docking analysis of these compounds showed that majority of the similar compounds were able to block the active site of PL^{pro} with either increased or similar binding efficiency, as indicated by their binding energy values (**Supplementary Table S3 B**). Except compound-507 (0.48 kcal/mol), a similar compound of cinacalcet and compound-509 (0.48 kcal/mol), a similar compound of procainamide, all others were able to bind efficiently to the active site of PL^{pro}. Me-too compounds with a better or similar binding energy can be found in **Supplementary Table S3.** A complete list of the compounds docked against PL^{pro} and their binding energy can be found in the Supplementary excel file. Out of these potential compounds we shortlisted two top hit lead compounds, compound-502 (-8.97 kcal/mol) and compound-565 (-6.94 kcal/mol), based on the binding energy difference compared to their corresponding reference compounds, praziquantel (-7.68 kcal/mol) and ribavirin (-5.71 kcal/mol), respectively (Fig. 5).

Praziquantel is an FDA-approved trematodicide drug, which is an anthelmintic agent. It is widely used for the treatment of worm infections [62]. CSNAP cheminformatics analysis of praziquantel resulted in five similar compounds (Fig. 5a). Among them, compound-502 was considered as the lead molecule as it showed the highest docking score (Scheme 4, Fig. 5c). A higher score can be explained by the fact that compound-502 forms three alkyl interactions with the amino acid residue LEU162 of PL^{pro} (Fig. 6a, b, Table 2), In contrast, LEU162 forms only one alkyl interaction with praziquantel (**Supplementary Figure S2A, S2B, Supplementary Table S4**). Additionally, compound-502 forms carbon–hydrogen bond interaction with TYR268, which further improves its binding efficiency. Moreover, activity and safety efficacy of this molecule is established against *Schistosoma mansoni* infections [63] and therefore maybe considered for further studies against COVID-19.

Ribavirin is a nucleoside analog, which is effective against the hepatitis C virus (HCV), human respiratory syncytial virus (RSV), and other wide range of RNA viruses [56]. It gets incorporated into viral RNA thereby disturbing viral RNA synthesis [64]. Ribavirin is one of the safest and most effective medicines, which is listed as an essential medicine by the WHO [65]. Recent studies have suggested that ribavirin may be effective against the treatment of COVID-19 [66]. Additionally, another in vitro study with Vero E6 cells showed that ribavirin inhibited SARS-CoV-2

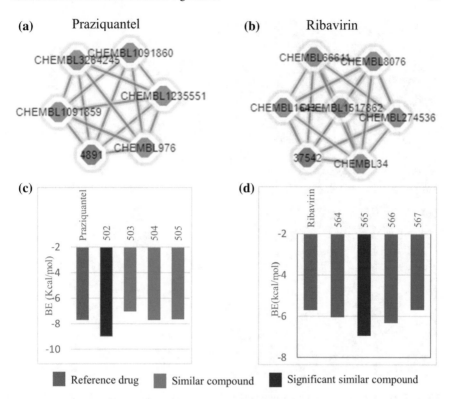

Fig. 5 Chemical similarity analysis using CSNAP to identify Me-too drugs against SARS-CoV-2 PL^{Pro}. Similarity network map for reference repurposed drugs **a** praziquantel, **b** ribavirin. **c–d** Column plots representing molecular docking score of the Me-too drugs compared to the corresponding reference drugs. (*Note*: Ligands that could not be docked due to structural complexity are not included in the bar graph.)

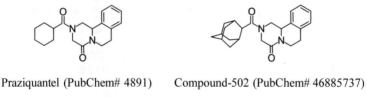

Praziquantel (PubChem# 4891)   Compound-502 (PubChem# 46885737)

Scheme 4 Chemical structure of praziquantel and compound-502

infection with an IC50 of 0.8 µM [67]. It is also shown to shorten the duration of virus shedding and decrease cytokine responses during the phase 2 trial [68]. Our CSNAP cheminformatics analysis of ribavirin resulted in six similar compounds (Fig. 5B). Upon molecular docking analysis, compound-565 showed the highest binding energy difference when compared to ribavirin (Scheme 5, Fig. 5d) compound-565, also known as levovirin, is an analog, and stereoisomer of ribavirin [69]. Upon analyzing

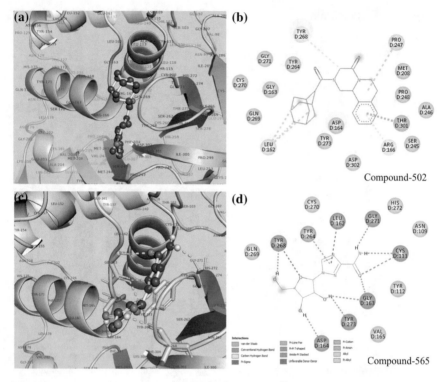

Fig. 6 Low-energy binding conformations of SARS-CoV-2 PL^Pro^–ligand docked complex showing 3D interactions (**a, c**) created by Pymol and 2D interactions (**b, d**) of the ligand with amino acid residues of PL^Pro^ created by discovery studio for compounds-502 and 565, respectively

Table 2 List of interacting amino acid residues of SARS-CoV-2 PL^Pro^ with selected novel lead compounds-502 and 565

Compound	PubChem ID	Hydrogen bonds	Hydrophobic interaction	Other interactions	Binding energy (kcal/mol)
502	46,885,737	–	LEU162, ARG166	TYR268, PRO247, THR301, ARG166, ALA246, ASP164, ASP273, SER245	−8.97
565	460,516	CYS111, GLY271, GLY163, TYR273, ASP164, TYR268	LEU162, TYR264	HIS272, TYR112, VAL165, CYS270, GLN269	−6.94

Ribavirin (PubChem# 37542)   Compound-565 (PubChem# 460516)

Scheme 5 Chemical structure of ribavirin and compound-565

the 3D and 2D interaction maps between the ligand and PL^{pro}, it was found that compound-565 interacts with the protein via eight hydrogen bonds; two each with CYS111 and GLY163, and one each GLY271, GLY163, TYR273, and ASP164 residues (Fig. 6c, d and Table 2). In contrast, ribavirin interacts with only six such bonds **(Supplementary Figure S5 C and S5 D).** Earlier reports on compound-565 show that it is has an antiviral effect against bovine viral diarrhea virus by inhibiting IMPDH enzyme [70]. Further, it is reported to show reduced toxicity, antiviral effects with retention of both immunomodulatory activity and reduction of hepatitis [69, 71]. With our results showing improved binding affinity of this molecule for PL^{pro}, it may be considered as a better therapeutic candidate than ribavirin against SARS-CoV-2.

We further confirmed their binding by crosschecking with the ribbon structures of protein and confirmed that all of these ligands docked at the active site of the protease enzyme (Fig. 7).

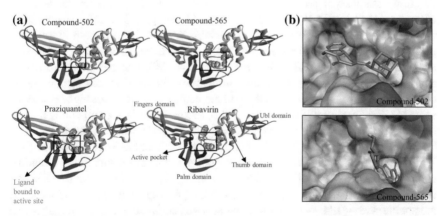

Fig. 7 Docked protein complex of top hit compounds for SARS-Cov-2 PL^{Pro} protein. **a** Ribbon structure of docked protein complex of top hit compounds-502 and 565 with their reference drugs, praziquantel and ribavirin, showing ligands docked at the active pocket of the protease (highlighted in square box). **b** Surface protein image and active site binding of compounds-502 and 565 surface colored by hydrogen bond type

3.4   Inhibitors Against RNA-Dependent RNA Polymerase

RNA-dependent RNA polymerase (RdRP), also known as NSP12, is a key component of the replication–transcription complex of SARS-CoV-2. It plays a central role in the viral replication and transcription cycle in association with two other NSPs, NSP7 and NSP8. It is therefore, considered as an ideal target for drug designing. During the preparation of this manuscript, reports show that more than 24 experimental drugs are repurposed for targeting RdRP and some are under clinical trials. These include cortisone, novobiocin, silybin, saquinavir, tipranavir, lonafarnib, filibuvir, simprevir, cepharanthine, remdesivir, galidesivir, tenofovir, sofosbuvir, ribavirin, setrobuvir, balapiravir, mericitabine, IDX-184, BMS-986094, YAK, PSI-6130, R-1479, chlorohexidine, and chenodeoxycholic acid [11, 13, 16–18, 56, 72–75] (**Supplementary Table S5 A**). Using CSNAP analysis with these drugs as the main reference compounds we identified a total of 152 similar compounds showing a similarity score of more than 0.85. A complete list of the compounds can be found in **Supplementary File A.** The binding energy of these similar ligands generated from molecular docking analysis shows that the Me-too compounds bind either similarly or more effectively to the active site of the enzyme compared to the respective reference molecules (**Supplementary File A, Supplementary Table S5 B**). It is worthwhile to mention that compound-155 (−7.47 kcal/mol), compound-124 (−8.81 kcal/mol), and compound-20 (−8.52 kcal/mol) showed improved binding energy compared to their corresponding reference compounds remdesivir (−5.24 kcal/mol), tipranavir (−6.93 kcal/mol), and novobiocin (−6.8 kcal/mol), respectively (Fig. 8).

Remdesivir has been reported to show promise for the treatment of COVID-19 [76, 77]. Initial studies with remdesivir showed that it was able to reduce the recovery time of hospitalized COVID-19 patients from 15 to 11 days [78]. Recently, accumulating multiple evidences show remdesivir can inhibit SARS-COV-2 in both in vitro and in vivo models [79]. Another in vivo study revealed that remdesivir can be an antiviral agent through targeting RdRP in rhesus monkeys [80]. CSNAP cheminformatics analysis of remdesivir resulted in nine similar compounds (Fig. 8a). Among them, compound-155 was considered as the lead molecule as it showed the highest docking score of −7.47 kcal/mol (Scheme 6, Fig. 8d). Furthermore, it is worth noting that the main plasma metabolite of remdesivir, is GS-441524 [81], which is shown to inhibit SARS-COV-2 infection in Vero E6 cells in vitro with an EC50 of 5.188 μM [82]. When this active metabolite was docked against RdRP, our results show a binding energy of −5.55 kcal/mol. It is therefore, imperative that compound-155 shows improved binding affinity compared to both the prodrugs, remdesivir and its active metabolite, GS-44124. A higher binding score of compound-155 can be explained by the fact that this molecule interacts with RdRP amino acid residues, TYR619, PHE793, and SER795, via three hydrogen bonds (Fig. 9a, b, Table 3), whereas remdesivir is involved in only two such interactions (**Supplementary Figure S3A, S3B, Supplementary Table S6**). Further, an extra benzyl group in compound-155 helps it to interact with the enzyme with additional hydrophobic and van der Waals

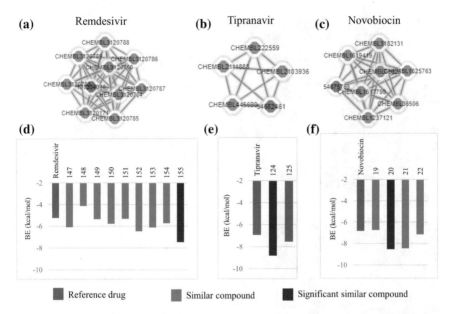

Fig. 8 Chemical similarity analysis using CSNAP to identify Me-too drugs against SARS-CoV-2 RdRP. Similarity network map for reference repurposed drugs **a** remdesivir, **b** tipranavir, and **c** novobiocin. **d–f** Column plots representing molecular docking score of the Me-too drugs compared to the corresponding reference drugs. (*Note*: Ligands that could not be docked due to structural complexity are not included in the bar graph.)

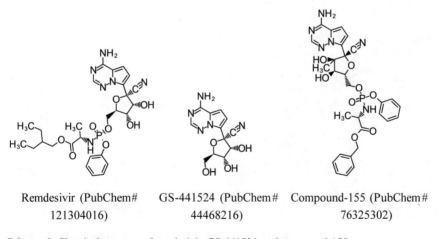

Remdesivir (PubChem# 121304016)

GS-441524 (PubChem# 44468216)

Compound-155 (PubChem# 76325302)

Scheme 6 Chemical structure of remdesivir, GS-441524, and compound-155

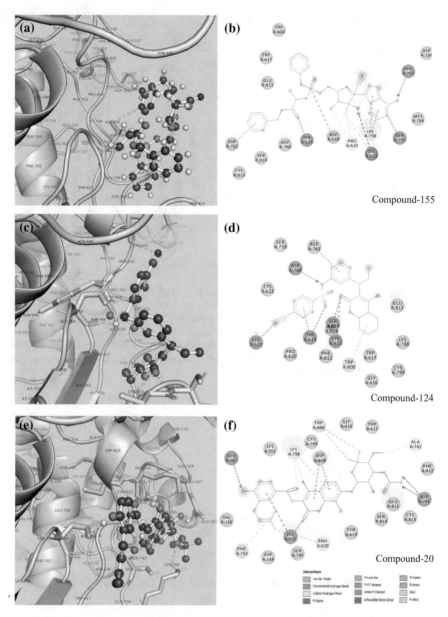

Fig. 9 Low-energy binding conformations of SARS-CoV-2 RdRP-ligand docked complex. Showing 3D interactions (**a, c, e**) created by Pymol, 2D interaction (**b, d, f**) of ligand with amino acid residues of RdRP created by discovery studio for compounds-155, 124, and 20 respectively

Table 3 List of interacting amino acid residues of SARS-CoV-2 RdRP with selected novel lead compounds-155, 124, and 20

Compound entry no	PubChem ID	Hydrogen bonds	Hydrophobic interaction	Other interactions	Binding energy (kcal/mol)
155	76325302	TYR619, PHE793, SER795	LYS798	ASP618, LYS798, PRO620, LYS621	−7.47
124	54685540, 163329584	LYS621, TYR619, ASP760 (2), CYS813, SER814	TYR619, ASP761, TRP800	CYS622, PRO620, PHE812, ASP618, GLU811, TRP617, CYS799, GLY616	−8.81
20	54713167	GLU167 , LYS621, ASP761	PHE793, LYS621, PRO620, ASP618, LYS798, TRP800	PRO620, LYS798, ALA762, LYS551, CYS799	−8.52

interactions with PRO620, LYS621, and LYS798 (Fig. 9b). Interestingly, compound-155 is also reported to be an effective antiviral drug with activity against HCV [83]. Apart from the binding affinity for RdRP and potential antiviral activity, this molecule may be further explored as a potential drug against SARS-CoV-2.

Tipranavir is widely used in antiretroviral therapy (ART) for the treatment of acquired immunodeficiency syndrome (AIDS) patients [84]. It has potent antiretroviral protease inhibition activity and is shown to inhibit human immunodeficient virus (HIV) [85]. Tipranavir binds to the active site of the HIV protease, thereby inhibiting the activity of enzyme. This inhibition prevents the cleavage of the polyprotein which results in immature non-infectious viral particles. Recent studies show that tipranavir was able to inhibit SARS-CoV-2 replication in vitro in VeroE6 cells [72]. CSNAP cheminformatics analysis of tipranavir resulted in four similar compounds (Fig. 8b). Among them, compound-124 was considered as the lead molecule because it showed the highest docking score (Scheme 7, Fig. 8e). Upon analyzing the 3D and 2D interaction maps between the ligand and RdRP, it was found that compound-124 forms a total six hydrogen bonds, one with each amino acid residues LYS621, TYR619, CYS813, and SER814 additionally two bonds with ASP760 (Fig. 9c, d, Table 3). In contrast, tripranavir interacts with the protein with only three hydrogen bonds **(Supplementary Figure S3A, S3B, Supplementary Table S6)**.

Novobiocin is an aminocoumarin group of antibiotic drug. It is known to be a potent inhibitor of bacterial DNA gyrase and ATPase [86]. It has also shown anti-viral activity against zika and vaccinia viruses earlier [87, 88]. CSNAP cheminformatics

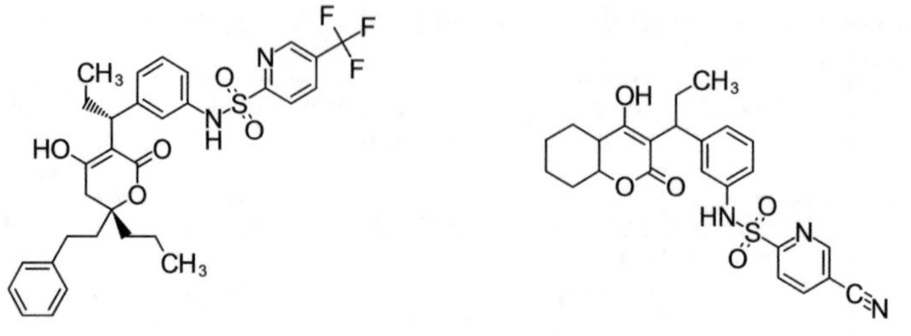

Tipranavir (PubChem# 4682461)   Compound-124 (PubChem#54685540)

Scheme 7 Chemical structure of tipranavir and compound-124

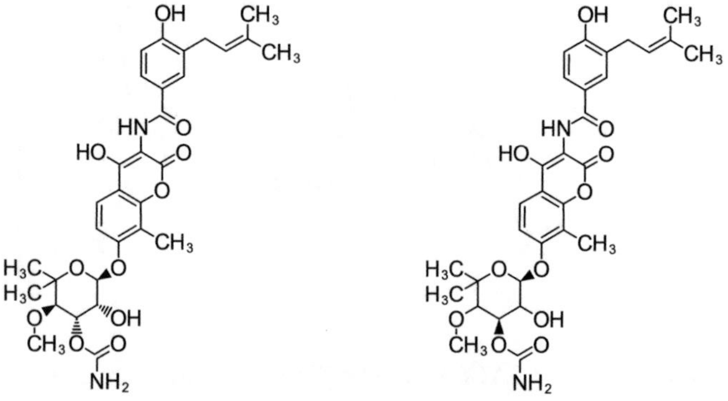

Novobiocin (PubChem# 54675769)   Compound-20 (PubChem# 54713167)

Scheme 8 Chemical structure of novobiocin and compound-20

analysis of novobiocin resulted in seven similar compounds (Fig. 8c). Among them, compound-20 was considered as the lead molecule because it showed the highest docking score (Scheme 8, Fig. 8f). Upon analyzing the 3D and 2D interaction maps between the ligand and RdRP, it was found that this molecule forms four hydrogen bonds, one each with amino acid residues, GLU167, LYS621, and two bonds with ASP761 (Fig. 9e, f, Table 3). In contrast, novobiocin interacts with RdRP protein with only three hydrogen bonds **(Supplementary Figure S3E, S3F, Supplementary Table S6)**. Additionally, compound-20 was able to form the carbon–hydrogen bond interactions with ALA762, LYS798, and PRO620. All of these lead ligands are verified manually for their binding site to confirm that they are binding to the active site of SARS-Cov-2 RdRP by comparing with the PDB structure 7BV2 which is an RdRP complex bound to remdesivir [89] (Fig. 10).

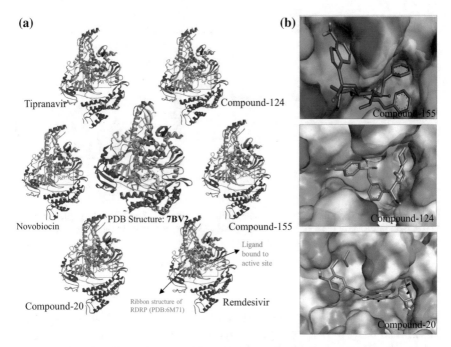

Fig. 10 Docked protein complex of top hit compounds for SARS-Cov-2 RdRP protein. **a** ribbon structure of complex compared with the PDB structure remdesivir–RdRP complex 7BV2 (nsp12–nsp7–nsp8 complex bound to the template-primer RNA and remdesivir). All the ligands docked at the active pocket (highlighted in square box) of RdRP similar to 7BV2. **b** Surface protein image and active site binding of compound-155, compound-124, and compound-20, surface colored by hydrogen bond type

3.5 Molecular Dynamic Simulation

Among the eight best Me-too molecules, compound-155 showed the highest binding energy difference compared to its corresponding reference molecule, remdesivir. We, therefore, selected it for molecular dynamics simulation analysis to assess its binding stability with RdRP. Our simulation analysis shows that the ligand remained close to its original position with an RMSD of 3 Å up to 10 ns. Between 10 and 50 ns, the ligand slightly deviated compared to the protein. Toward the end of the simulation, between 50 and 70 ns, the ligand returned close to its original position with an RMSD of 3 Å (Fig. 11a). These results indicate that compound-155 is stable in binding to the RdRP protein. To assess the binding affinity, MM-PBSA analysis was performed to calculate the binding energy, which was found to be −67.778 kcal/mol, indicating strong interaction of compound-155 with RdRP. Further, our hydrogen bond analysis indicates that ASP618, LYS798, and GLU811 residues are involved in hydrogen bond interaction with the protein in majority of the time frames during the simulation run (Fig. 11b). These residues impart stability to the protein–ligand complex.

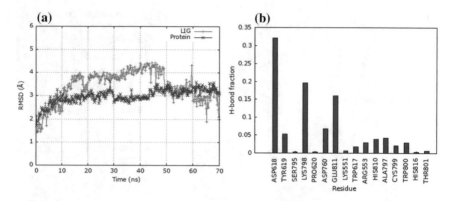

Fig. 11 Molecular dynamics simulation analysis of compound-155 and RdRP complex. **a** 70 ns-MD simulation RMSD plots of ligand (compound-155) bound RdRP. **b** Hydrogen bond fraction interaction with the selected amino acid residues

4   Conclusion

The ongoing pandemic of COVID-19 has become a threat to human health and life with no drugs or other therapeutics presently approved by the FDA. Since the onset of the disease, accumulating evidences reveal that trials for several promising drugs against the disease were either halted due to safety concerns or completed with showing poor and limited efficacy [20–23]. Furthermore, vaccines alone cannot eliminate the COVID-19 pandemic. It is now widely recognized that a collective approach including widespread vaccination and therapeutics will be required to control the pandemic. Hence, a constant search for novel therapeutics is essential. It is therefore, imperative to explore more potential drugs to combat this pandemic. In the current study, we have used a chemical similarity-based approach to identify more than 450 potential ligands including Me-too drugs that are similar to drugs being explored for SARS-CoV-2. Upon molecular docking analysis, we have shortlisted three lead molecules each for 3CL^pro and RdRP and two for PL^pro based on their binding energy values. These lead molecules show better interactions compared to the reference molecules due to an increase in hydrogen bonds or other interactions with the viral proteins. Finally, among the eight lead molecules, we selected compound-155 for molecular dynamics simulation analysis. Our results indicate that the protein–ligand complex for this molecule is stable during the 70 ns run. In summary, our *in-silico* findings identify several molecules, can be potential drugs inhibiting different stages of the SARS-CoV-2 viral lifecycle (Fig. 12), or can be used as a new way for developing novel drugs against the disease. However, further in vitro and in vivo studies are essential to ascertain our findings.

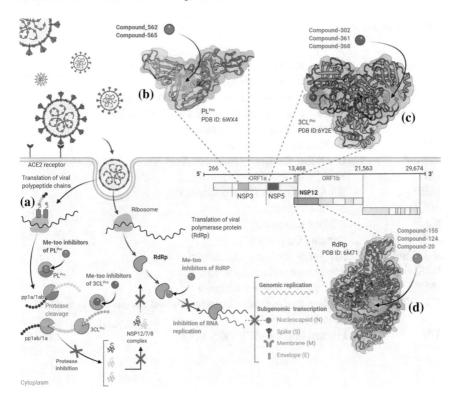

Fig. 12 **a** Early stages of SARS-CoV-2 lifecycle, early events of protease cleavage by PL^{Pro}, and 3CL^{Pro} to yield NSPs from pp1a and pp1ab. Second stage where RdRP translates viral RNA to multiple viral proteins. **b** Structure of PL^{Pro} and its active site (PDB structure-6WX4). **c** Structure of 3CL^{Pro} and its active site (PDB structure-6Y2E). **d** Structure of RdRP and its active site (PDB structure-6M71) where lead Me-too inhibitors bind

Acknowledgments We would like to acknowledge https://biorender.com which was used to create Fig. 12. Also, Ms. Ramya H is acknowledged for providing critical reviews and editing the manuscript.

Author Contributions MD, MNJ, and JPU conceived and designed the experiments; JPU carried out CSNAP analysis and screening; MNJ, MM, and KS carried out docking experiments; EI carried out molecular dynamics simulations; MNJ, JPU, MM, and MD performed data analysis; MNJ, JPU, EI, and MD wrote the manuscript; EI and MD provided critical inputs to the manuscript. MD supervised overall project. All authors contributed to manuscript revision, read, and have given approval to the final version of the manuscript.

Conflict of Interest All authors declare no conflict of interest.

Funding Sources This research did not receive any specific grant from funding agencies in the public, commercial, or not-for-profit sectors.

Abbreviations

SARS-CoV-2	Severe acute respiratory syndrome coronavirus 2;
COVID-19	Novel 2019 Coronavirus Disease;
SARS-CoV	Severe acute respiratory syndrome coronavirus;
MERS-CoV	Middle East respiratory syndrome coronavirus;
FDA	Food and Drug Administration;
WHO	World Health Organization;
CSNAP	Chemical Similarity Network Analysis Pull-down;
NSP	Non-Structural Protein;
RdRP	RNA-dependent RNA polymerase;
M^{pro}	Main protease;
$3CL^{pro}$	3C-like protease;
PL^{pro}	Papain-like protease;
MD simulations	Molecular dynamic simulation;
RMSD	Root mean square deviation;
RMSF	Root mean square fluctuation;
MM-PBSA	Molecular mechanics Poisson-Boltzmann Surface Area;
HCV	Hepatitis C virus;
HIV	Human immunodeficiency virus

References

1. Gorbalenya AE et al (2020) The species Severe acute respiratory syndrome-related coronavirus: classifying 2019-nCoV and naming it SARS-CoV-2. Nat Microbiol 5(4):536–544
2. Cucinotta D, Vanelli M (2020) WHO declares COVID-19 a pandemic. Acta Bio Medica: Atenei Parmensis 91(1):157
3. WHO, COVID-19 Dashboard Geneva: World Health Organization, 2020. https://covid19.who.int/ (Accessed 30 Aug 2021).
4. Jabbari P, Rezaei N (2020) With risk of reinfection, is COVID-19 here to stay? Disaster Med Public Health Prep 14(4):e33–e33
5. Tillett RL et al (2021) Genomic evidence for reinfection with SARS-CoV-2: a case study. Lancet Infect Dis 21(1):52–58
6. Tortorici MA, Veesler D (2019) Chapter four—structural insights into coronavirus entry. In: Rey FA (ed) Advances in virus research. Academic Press, pp 93–116
7. Surjit M, Lal SK (2009) The nucleocapsid protein of the sars coronavirus: structure, function and therapeutic potential. In: Molecular biology of the SARS-coronavirus, pp 129–151
8. Narayanan K et al (2000) Characterization of the coronavirus M protein and nucleocapsid interaction in infected cells. J Virol 74(17):8127–8134
9. Opstelten DJ et al (1995) Envelope glycoprotein interactions in coronavirus assembly. J Cell Biol 131(2):339–349
10. Fehr AR, Perlman S (2015) Coronaviruses: an overview of their replication and pathogenesis. Methods Mol Biol (Clifton NJ) 1282:1–23
11. Tsuji M (2020) Potential anti-SARS-CoV-2 drug candidates identified through virtual screening of the ChEMBL database for compounds that target the main coronavirus protease. FEBS Open Bio 10(6):995–1004

12. Torequl Islam M, et al (2020) A perspective on emerging therapeutic interventions for COVID-19. Front Public Health 8(281)
13. Elfiky AA (2021) SARS-CoV-2 RNA dependent RNA polymerase (RdRp) targeting: an in silico perspective. J Biomol Struct Dyn 39(9):3204–3212
14. Chen YW, Yiu C-PB, Wong K-Y (2020) Prediction of the SARS-CoV-2 (2019-nCoV) 3C-like protease (3CL (pro)) structure: virtual screening reveals velpatasvir, ledipasvir, and other drug repurposing candidates. F1000Research 9:129
15. Chtita S et al (2021) Discovery of potent SARS-CoV-2 inhibitors from approved antiviral drugs via docking and virtual screening. Comb Chem High Throughput Screen 24(3):441–454
16. Joshi RS et al (2021) Discovery of potential multi-target-directed ligands by targeting host-specific SARS-CoV-2 structurally conserved main protease. J Biomol Struct Dyn 39(9):3099–3114
17. Wu C et al (2020) Analysis of therapeutic targets for SARS-CoV-2 and discovery of potential drugs by computational methods. Acta Pharmaceutica Sinica B 10(5):766–788
18. Narkhede RR et al (2020) The molecular docking study of potential drug candidates showing anti-COVID-19 activity by exploring of therapeutic targets of SARS-CoV-2. Eurasian J Med Oncol 4(3):185–195
19. Saha RP, et al (2020) Repurposing drugs, ongoing vaccine, and new therapeutic development initiatives against COVID-19. Front Pharmacol 11(1258)
20. Elkin ME, Zhu X (2021) Understanding and predicting COVID-19 clinical trial completion versus cessation. Plos One 16(7):e0253789
21. Lee Z et al (2021) The rise and fall of hydroxychloroquine for the treatment and prevention of COVID-19. Am J Trop Med Hyg 104(1):35
22. Cao B et al (2020) A trial of lopinavir-ritonavir in adults hospitalized with severe covid-19. N Engl J Med 382(19):1787–1799
23. Consortium WST (2021) Repurposed antiviral drugs for COVID-19—interim WHO SOLIDARITY trial results. N Engl J Med 384(6):497–511
24. Lo YC, et al (2015) Large-scale chemical similarity networks for target profiling of compounds identified in cell-based chemical screens. PLoS Comput Biol 11(3):e1004153
25. Lo Y-C et al (2018) Machine learning in chemoinformatics and drug discovery. Drug Discov Today 23(8):1538–1546
26. Lo Y-C et al (2016) 3D chemical similarity networks for structure-based target prediction and scaffold hopping. ACS Chem Biol 11(8):2244–2253
27. Aronson JK, Green AR (2020) Me-too pharmaceutical products: history, definitions, examples, and relevance to drug shortages and essential medicines lists. Br J Clin Pharmacol 86(11):2114–2122
28. Baldi P, Benz RW (2008) BLASTing small molecules—statistics and extreme statistics of chemical similarity scores. Bioinformatics 24(13):i357–i365
29. Zhang L et al (2020) Crystal structure of SARS-CoV-2 main protease provides a basis for design of improved α-ketoamide inhibitors. Science 368(6489):409–412
30. Rut W, et al (2020) Activity profiling and crystal structures of inhibitor-bound SARS-CoV-2 papain-like protease: a framework for anti–COVID-19 drug design. Sci Adv 6(42):eabd4596
31. Gao Y et al (2020) Structure of the RNA-dependent RNA polymerase from COVID-19 virus. Science 368(6492):779–782
32. Pettersen EF et al (2004) UCSF Chimera—a visualization system for exploratory research and analysis. J Comput Chem 25(13):1605–1612
33. Hanwell MD et al (2012) Avogadro: an advanced semantic chemical editor, visualization, and analysis platform. J Cheminformatics 4(1):17
34. Shannon A, et al (2020) Remdesivir and SARS-CoV-2: structural requirements at both nsp12 RdRp and nsp14 Exonuclease active-sites. Antivir Res 104793
35. Mirza MU, Froeyen M (2020) Structural elucidation of SARS-CoV-2 vital proteins: computational methods reveal potential drug candidates against main protease, Nsp12 polymerase and Nsp13 helicase. J Pharm Anal 10(4):320–328

36. Berendsen HJ, van der Spoel D, van Drunen R (1995) GROMACS: a message-passing parallel molecular dynamics implementation. Comput Phys Commun 91(1–3):43–56
37. Lee J et al (2016) CHARMM-GUI input generator for NAMD, GROMACS, AMBER, OpenMM, and CHARMM/OpenMM simulations using the CHARMM36 additive force field. J Chem Theory Comput 12(1):405–413
38. Jorgensen WL et al (1983) Comparison of simple potential functions for simulating liquid water. J Chem Phys 79(2):926–935
39. Chen F et al (2004) In vitro susceptibility of 10 clinical isolates of SARS coronavirus to selected antiviral compounds. J Clin Virol 31(1):69–75
40. Ou X et al (2020) Characterization of spike glycoprotein of SARS-CoV-2 on virus entry and its immune cross-reactivity with SARS-CoV. Nat Commun 11(1):1–12
41. Shang J et al (2020) Cell entry mechanisms of SARS-CoV-2. Proc Natl Acad Sci 117(21):11727–11734
42. Zhou Y et al (2020) Network-based drug repurposing for novel coronavirus 2019-nCoV/SARS-CoV-2. Cell Discov 6(1):1–18
43. Elfiky AA (2020) Ribavirin, Remdesivir, Sofosbuvir, Galidesivir, and Tenofovir against SARS-CoV-2 RNA dependent RNA polymerase (RdRp): a molecular docking study. Life Sci 117592
44. McConathy J, Owens MJ (2003) Stereochemistry in drug action. Prim Care Companion J Clin Psychiatry 5(2):70–73
45. Bouhaddou M, et al (2020) The global phosphorylation landscape of SARS-CoV-2 infection. Cell 182(3): 685–712. e19
46. Mukundan Satyanarayanan DV (2020) Ligands based drug design for covid 19-A multi-faceted approach using ligand design, molecular docking and binding probability calculation. Int J Res Appl Sci & Eng Technol (IJRASET) 8:844–850
47. Suravajhala R et al (2021) Molecular docking and dynamics studies of curcumin with COVID-19 proteins. Netw Model Anal Health Inform Bioinform 10(1):44
48. Das S et al (2021) An investigation into the identification of potential inhibitors of SARS-CoV-2 main protease using molecular docking study. J Biomol Struct Dyn 39(9):3347–3357
49. Labrie F (1988) Combination therapy for treatment of female breast cancer. https://patents.goo gle.com/patent/US4775660A/en (Accessed on 30 Aug 2021). 1988, Google Patents
50. Kim K-J et al (2020) PI3K-targeting strategy using alpelisib to enhance the antitumor effect of paclitaxel in human gastric cancer. Sci Rep 10(1):1–12
51. Markham A (2019) Alpelisib: first global approval. Drugs 79(11):1249–1253
52. André F et al (2019) Alpelisib for PIK3CA-mutated, hormone receptor–positive advanced breast cancer. N Engl J Med 380(20):1929–1940
53. Rothman J (2017) The rediscovery of bisantrene: a review of the literature. Int J Cancer Res Ther 2:1–10
54. Sun L, et al (2012) Coronavirus papain-like proteases negatively regulate antiviral innate immune response through disruption of STING-mediated signaling. PloS One 7(2):e30802
55. Arya R, et al (2020) Potential inhibitors against papain-like protease of novel coronavirus (SARS-CoV-2) from FDA approved drugs. https://doi.org/10.26434/chemrxiv.118 60011.ChemRxiv
56. Elfiky AA (2020) Anti-HCV, nucleotide inhibitors, repurposing against COVID-19. Life Sci117477
57. Estrada E (2020) COVID-19 and SARS-CoV-2. Modeling the present, looking at the future. Phys Rep 869:1–51
58. Singh YD, et al (2020) Potential bioactive molecules from natural products to combat against coronavirus. Adv Tradit Med 1–12
59. Yu R, et al (2020) Computational screening of antagonists against the SARS-CoV-2 (COVID-19) coronavirus by molecular docking. Int J Antimicrob Agents 56(2):106012
60. Amin SA, et al (2020) Protease targeted COVID-19 drug discovery and its challenges: insight into viral main protease (Mpro) and papain-like protease (PLpro) inhibitors. Bioorganic Med Chem 115860

61. Xu Z, et al (2020) Nelfinavir was predicted to be a potential inhibitor of 2019-nCov main protease by an integrative approach combining homology modelling, molecular docking and binding free energy calculation, 921627. 27 Jan 2020. https://doi.org/10.1101/2020.01.27.921627

62. Park S-K et al (2019) The anthelmintic drug praziquantel activates a schistosome transient receptor potential channel. J Biol Chem 294(49):18873–18880

63. Dong Y et al (2010) Praziquantel analogs with activity against juvenile Schistosoma mansoni. Bioorg Med Chem Lett 20(8):2481–2484

64. Te HS, Randall G, Jensen DM (2007) Mechanism of action of ribavirin in the treatment of chronic hepatitis C. Gastroenterol Hepatol 3(3):218

65. WHO (2019) WHO Model Lists of Essential Medicines, 21st List 2019. Accessed on 30 Aug 2021. https://www.who.int/publications/i/item/WHOMVPEMPIAU2019.06

66. Khalili JS et al (2020) Novel coronavirus treatment with ribavirin: groundwork for an evaluation concerning COVID-19. J Med Virol 92(7):740–746

67. Unal MA et al (2021) Ribavirin shows antiviral activity against SARS-CoV-2 and downregulates the activity of TMPRSS2 and the expression of ACE2 in vitro. Can J Physiol Pharmacol 99(5):449–460

68. Hung IF-N et al (2020) Triple combination of interferon beta-1b, lopinavir–ritonavir, and ribavirin in the treatment of patients admitted to hospital with COVID-19: an open-label, randomised, phase 2 trial. Lancet 395(10238):1695–1704

69. Fang C, Srivastava P, Lin CC (2003) Effect of ribavirin, levovirin and viramidine on liver toxicological gene expression in rats. J Appl Toxicol: Int J 23(6):453–459

70. Stuyver LJ et al (2002) Inhibitors of the IMPDH enzyme as potential anti-bovine viral diarrhoea virus agents. Antiviral Chem Chemother 13(6):345–352

71. Tam RC et al (2000) The ribavirin analog ICN 17261 demonstrates reduced toxicity and antiviral effects with retention of both immunomodulatory activity and reduction of hepatitis-induced serum alanine aminotransferase levels. Antimicrob Agents Chemother 44(5):1276–1283

72. Yamamoto N, et al (2020) Nelfinavir inhibits replication of severe acute respiratory syndrome coronavirus 2 in vitro, BioRxiv. https://doi.org/10.1101/2020.04.06.026476

73. Ruan Z et al (2021) SARS-CoV-2 and SARS-CoV: virtual screening of potential inhibitors targeting RNA-dependent RNA polymerase activity (NSP12). J Med Virol 93(1):389–400

74. Indu P et al (2020) Raltegravir, Indinavir, Tipranavir, Dolutegravir, and Etravirine against main protease and RNA-dependent RNA polymerase of SARS-CoV-2: A molecular docking and drug repurposing approach. J Infect Public Health 13(12):1856–1861

75. Choudhury S et al (2021) Evaluating the potential of different inhibitors on RNA-dependent RNA polymerase of severe acute respiratory syndrome coronavirus 2: a molecular modeling approach. Med J Armed Forces India 77:S373–S378

76. Grein J et al (2020) Compassionate use of remdesivir for patients with severe Covid-19. N Engl J Med 382(24):2327–2336

77. Gordon CJ et al (2020) The antiviral compound remdesivir potently inhibits RNA-dependent RNA polymerase from Middle East respiratory syndrome coronavirus. J Biol Chem 295(15):4773–4779

78. Beigel JH et al (2020) Remdesivir for the treatment of Covid-19. N Engl J Med 383(19):1813–1826

79. Frediansyah A et al (2021) Remdesivir and its antiviral activity against COVID-19: a systematic review. Clin Epidemiol Glob Health 9:123–127

80. Williamson BN et al (2020) Clinical benefit of remdesivir in rhesus macaques infected with SARS-CoV-2. Nature 585(7824):273–276

81. Yan VC, Muller FL (2020) Advantages of the parent nucleoside GS-441524 over remdesivir for Covid-19 treatment. ACS Med Chem Lett 11(7):1361–1366

82. Shi Y et al (2021) The preclinical inhibitor GS441524 in combination with GC376 efficaciously inhibited the proliferation of SARS-CoV-2 in the mouse respiratory tract. Emerg Microbes Infect 10(1):481–492

83. National Center for Biotechnology Information (2020) PubChem bioassay record for bioactivity AID 1070585—SID 194167708, B.f.A.-S., Source: ChEMBL. Accessed 30 Aug 2020. https://pubchem.ncbi.nlm.nih.gov/bioassay/1070585#sid=194167708

84. Hicks CB et al (2006) Durable efficacy of tipranavir-ritonavir in combination with an optimised background regimen of antiretroviral drugs for treatment-experienced HIV-1-infected patients at 48 weeks in the randomized evaluation of strategic intervention in multi-drug reSistant patients with Tipranavir (RESIST) studies: an analysis of combined data from two randomised open-label trials. Lancet 368(9534):466–475

85. Lv Z, Chu Y, Wang Y (2015) HIV protease inhibitors: a review of molecular selectivity and toxicity. HIV/AIDS (Auckland, NZ), 7:95–104

86. Khan T et al (2018) DNA gyrase inhibitors: progress and synthesis of potent compounds as antibacterial agents. Biomed Pharmacother 103:923–938

87. Yuan S et al (2017) Structure-based discovery of clinically approved drugs as Zika virus NS_2B-NS_3 protease inhibitors that potently inhibit Zika virus infection in vitro and in vivo. Antiviral Res 145:33–43

88. Sekiguchi J, Shuman S (1997) Novobiocin inhibits vaccinia virus replication by blocking virus assembly. Virology 235(1):129–137

89. Yin W et al (2020) Structural basis for inhibition of the RNA-dependent RNA polymerase from SARS-CoV-2 by remdesivir. Science 368(6498):1499–1504

Influences of Top-Surface Topography on Structural and Residual Trapping During Geological CO_2 Sequestration

Pradeep Reddy Punnam, Balaji Krishnamurthy,
and Vikranth Kumar Surasani

Abstract This work was conducted to investigate the influence of caprock topography parameters on structural trapping and residual trapping during geological CO_2 sequestration. The study is carried out on two types of geological folds, asymmetrical and chevron folds, which are integrated on three different sloping synthetic domains. Results show that the structural and residual trapping depends on the types of folds and perturbations on the geological domains. Geological parameters are analysed by evaluating the sweeping efficiency and entrapment percentage of the structural and residual trapping. The outcome of the study provides insight into the influence of the topography on geological CO_2 sequestration.

1 Introduction

The rise of CO_2 concentration in the atmosphere has been the major concern that contributing global warming. Significant CO_2 emissions are released into the atmosphere due to the anthropogenic activities of a human being. Industries from various sectors (mainly power and pyrometallurgy) are main contributors to CO_2 emissions. Carbon Capture and Storage (CCS) can be a considerable technology where the CO_2 emission into the atmosphere can be substantially reduced [1]. In the CCS technology, the geological CO_2 sequestration process is one of the critical processes where the captured CO_2 is injected into the geological subsurface, where the CO_2 is trapped and stored safely from entering the earth's atmosphere [2].

Due to in situ thermodynamic conditions, CO_2 is injected in supercritical state as it occupies less space than the gaseous state. Further in the manuscript, the supercritical CO_2 is referred to as CO_2. The injected CO_2 moves upwards in the subsurface due to the buoyancy. Once the CO_2 reaches the impermeable caprock, its upward movement is restricted and tends to migrate in the lateral direction depending on the caprock

P. R. Punnam · B. Krishnamurthy · V. K. Surasani (✉)
Department of Chemical Engineering, Birla Institute of Technology and Science,
Pilani-Hyderabad Campus, Hyderabad, India
e-mail: surasani@hyderabad.bits-pilani.ac.in

© The Author(s), under exclusive license to Springer Nature Singapore Pte Ltd. 2022   113
R. Srinivas et al. (eds.), *Advances in Computational Modeling and Simulation*,
Lecture Notes in Mechanical Engineering,
https://doi.org/10.1007/978-981-16-7857-8_9

topography. The CO_2 plume, which moves freely in the observable domain without any restraints, is called moving plume. During lateral migration, the CO_2 plume will get trap in the geological domain's top surface perturbations; this trapping mechanism is called structural trapping. When the CO_2 is migrating upwards from the injection point, it encounters numerous traps, restricting the CO_2 from migrating upwards. This quantity of CO_2 gets clogged in the migration pathway is called residual trapping [3–5].

The structurally and residually trapped CO_2 will interact with the connate water and dissolute into it; this trapping mechanism is called solubility trapping. The injected CO_2 that undergoes dissolution reaction with the connate water forms weak carbonic acids and decreases the pH surrounding domain, which will trigger the mineral reaction in the domain. This mineralisation that happens in the geological subsurface domain is known as the mineral trapping mechanism. This way, the harmful CO_2 is eradicated in the subsurface geological domain using the geological CO_2 sequestration process [1]. The structural and residual trapping mechanism is crucial among the four trapping mechanisms during the primary phase of the geological CO_2 sequestration process.

During geological CO_2 sequestration process, many geological parameters influence the structural and residual trapping. Caprock topography is one of the geological parameters that can influence the safe storage of CO_2 [4, 6, 7]. This study focuses on top-surface topographical parameters (Top-surface perturbations and morphological structure) integrated with two types of folds (asymmetrical and chevron folds). Many researchers developed numerical tools to elucidate and to explain the influences of different topographical surfaces on CO_2 geological sequestration [5, 8].

2 Theory

2.1 Multiphase Flow Equations

The multiphase mass conservation equations for CO_2 and water, and energy conservation equation are solved to analyse the influences of top-surface topography on CO_2 geological sequestration [5, 8]. In formulating the conservation equations, three assumptions were made. First, it is assumed that only two phases water and CO_2 present; second, mineral trapping is neglected in the domain i.e. no geochemical reactions will take place in the domain. Third assumption is that initially the computation domain is fully saturated with water. The multiphase flow equations consist of mass conversation equations for both water and CO_2, and the total energy equation are illustrated in Eqs. (1)–(3) [8, 9].

$$\frac{\partial}{\partial t}\left[\phi\left(S_l\rho_l X_w^l + S_g\rho_g X_w^g\right)\right]$$
$$+ \nabla \cdot \left[q_l\rho_l X_w^l + q_g\rho_g X_w^g - \phi\left(S_l D_l\rho_l \nabla X_w^l + S_g D_g\rho_g \nabla X_w^g\right)\right] = Q_w \quad (1)$$

$$\frac{\partial}{\partial t}\left[\phi\left(S_l\rho_l X_c^l + S_g\rho_g X_c^g\right)\right]$$
$$+ \nabla \cdot \left[q_l\rho_l X_c^l + q_g\rho_g X_c^g - \phi\left(S_l D_l\rho_l \nabla X_c^l + S_g D_g\rho_g \nabla X_c^g\right)\right] = Q_c \quad (2)$$

$$\frac{\partial}{\partial t}\left[\phi\left(S_l\rho_l U_l + S_g\rho_g U_g\right) + (1-\phi)\rho_r C_r T\right]$$
$$+ \nabla \cdot \left(q_l\rho_l H_l + q_g\rho_g H_g - \kappa\nabla T\right) = Q_e \quad (3)$$

Equations (1) and (2) represent the conservation of water and CO_2 respectively, and Eq. (3) is energy conservation equation. In Eqs. (1)–(3), the first term is associated with accumulation, the second term is with net property flux, and the third term denotes the source/sink terms (Q_w, Q_c, and Q_e). The superscript l and g represent the liquid and gases phases. The ϕ denotes the porosity of the domain. The terms $q_{l \, or \, g}$, $S_{l \, or \, g}$, and $D_{l \, or \, g}$ represent the Darcy velocity, phase saturation, and phase diffusivity coefficient, respectively. Additionally, $H_{l \, or \, g}$ represents the enthalpy, and $U_{l \, or \, g}$ represents the internal energy for the phases l and g. The terms ρ_r, c_r, T, and κ, represent the rock density, rock heat capacity, temperature, and thermal conductivity, respectively. The term X denotes the mole fraction of the species. In this study, only the multiphase flow is considered to analyse the structural and residual trapping adequately. Therefore, only two species are considered with predefined phases, which are liquid water (l) and supercritical CO_2 (g) [8, 9].

The Brooks-Corey relation is used to relate the capillary pressure to invading fluid saturation as

$$S_g = \begin{cases} (P_c/P_e)^{-n_b}, \, if \, P_c > P_e \\ 1, \, if \, P_c \leq P_e \end{cases} \quad (4)$$

The P_e and P_c terms represent the entry pressure and capillary pressure. To form a relation between the relative permeability and saturation, the Brooks-Corey-Mualem equation is utilised, which is illustrated in Eqs. (5) and (6) [9].

$$k_{r,l} = (S_l)^{n_1+n_2 n_3} \quad (5)$$

$$k_{r,g} = \left(1 - S_{e,l}\right)^{n_1}\left[1 - (S_l)^{n_2}\right]^{n_3} \quad (6)$$

The terms $k_{r,l \, or \, g}$ represent the relative permeability of liquid water and supercritical CO_2. The terms n_1, n_2, and n_3 are parameter constants; the value of n_1 is 1, n_2 is $1 + 1/n_b$, and n_3 is 2. The term n_b is related to the pore-size distribution; its range is from 0.2 to 5; in current simulation analysis, the value is taken as 2.5 [9].

2.2 Modelling Synthetic Domain

To explain the influences of top surface topography, a numerical simulation is conducted on the modelled synthetic simulation domain. The geological topography of the structural domain is constructed by using the membrane and trigonometric functions in MATLAB numerical tool. The membrane function accommodates the anticline and syncline structure on the surface. The sine and cosine functions are used to integrate the arbitrary perturbations and folds structure onto the domain so that it can act as a naturally available geological formation layer [5, 9]. The synthetic domain considered in this manuscript is modelled so that the sloping elevation shows dominance for large regions on the top surface. The reason for this consideration is to show the influence and dependences of caprock elevation on the migrated plume and trapping mechanisms.

Two types of folds integrated on the three different domains are considered to study the influences of top-surface perturbations on the structural and residual trapping. The asymmetric and chevron folds are integrated on the (i) plain sloping domain, (ii) sloping anticline domain, and (iii) sloping high perturbation domain (including both anticline and syncline). Figure 1 illustrates the three-dimensional grid structure of all the considered simulation domains. The estimation of structural and residual trapping is carried out based on the porosity ϕ, pore-volume (V_s and V_r structural and residual trapping cells volume), and CO_2 saturation (S_{co_2}) of the cells. Following Eqs. (7) and (8) are used to estimate the entrapment percentage of structural and residual trapping. The remaining quantity of CO_2 in the domain apart from structural and residual trapping quantity is measured as movable plume [8].

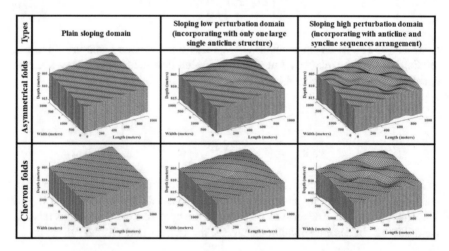

Fig. 1 Illustration of the three-dimensional grid structure of all the geological domains considered for structural and residual trapping

$$Structural\,trapping = \sum_{n=1}^{nf}\left(\varnothing V_s \rho_{co_2}\right) \times S_{co_2} \qquad (7)$$

$$Residual\,trapping = \sum_{n=1}^{nf}\left(\varnothing V_r \rho_{co_2}\right) \times S_{co_2} \qquad (8)$$

For the modelling of the synthetic computational domain, first, the top surface layer is constructed. Then the grid cells are assigned below the top surface layer to build the computation domain. The top surface layer is modelled by using the mesh grid function of MATLAB. The 50 grid cells are assigned in length and width direction, and in the depth axis, 20 grid cells are assigned. A total of 50,000 (50 × 50 × 20) grid cells are used to model the geological synthetic simulation domain. The quadratic cells are used in the modelling of the synthetic simulation domain. The physical dimensions of the synthetic simulation domain are 1000 × 1000 × 15 m. Each grid cells have eight vertices; these vertices are shared with neighbouring cells, except the top, bottom, and side face cells of a domain. The vertices coordinate in the width and length directions is assigned according to the physical dimensions of the domain. But the depth coordinates start from 800 m, which indicates that the CO_2 injection is carried out below 800 m. The pressure at each cell is calculated with using $\rho_w gh$. ρ_w represents the density of water, g represents the gravitational acceleration, and h is the depth axis cell centroid value [9, 10]. The pressure value for each individual cell changes according to the depth axis cell centroid of individual cells. The porosity range that is considered for all the domains is ranging from 0.2 to 0.4 [11]. The range of permeability of the domains is varying from 10 to 1500 mD [11]. The petrophysical properties for individual grid cells are arbitrarily assigned. The injection point for all the domains is selected at (240, 240, 810) coordinates. The volumetric injection rate of 1 m^3/day is considered for all the simulations of different synthetic domains.

3   Results

3.1   Influences of Top-Surface on Sweeping Efficiency

The results show that the chevron folds sweeping efficiency was slightly high compared to the asymmetrical folds in all considered synthetic domains. This set of observations has provided insight into the influences of geological folds on the sweeping efficiency. From Fig. 2, it can be notice that the sweeping efficiency in the anticline dome (third column) of the same fold is higher than the sloping plain domain (second column) results. It shows that the presence of topographical structure like anticline dome has a sure impact on the migration movement and sweeping efficiency. The anticline structure has increased the sweeping efficiency in the synthetic

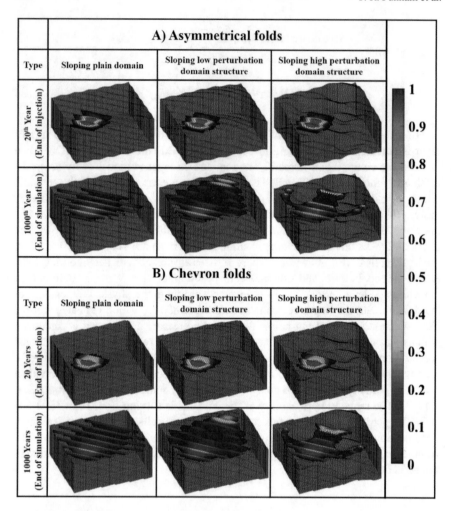

Fig. 2 Illustration of CO_2 saturation distribution in all three synthetic modelled geological domains, for **a** Asymmetrical folds and **b** Chevron folds. Results are presented for two discrete geological times, first at the end of the injection period (20 years) and second at the end of the simulation (1000 years)

geological domain. In the high perturbations integrated domain (fourth column) the migration path of CO_2 in the domain changes. The sweeping area covered by CO_2 plume was less compared to two other domains (plain domain and low perturbed domain integrated with anticline domain). In highly perturbed domain, the CO_2 plume movement is restricted in the domain. These results have shown that when the CO_2 has to be restricted under certain limits of the geological domain, injecting near the vicinity of the highly perturbed region is a good option. This will prevent the injected CO_2 plume from exploring the geological faults.

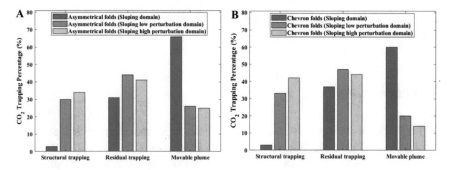

Fig. 3 Illustration of CO_2 entrapment percentage of structural trapping, residual trapping, and movable plume in all three synthetic domains at the 1000th year, for **a** Asymmetrical folds and **b** Chevron folds

3.2 Influences of Top-Surface on Structural and Residual Trapping Percentage

The previous section has illustrated and explained the sweeping efficiency of each domain that are used in this study. In Fig. 3, the histogram plots illustrate the structural trapping, residual trapping, and movable plume percentage for all synthetic domains obtained at the end of 1000 years.

As illustrated in Fig. 2, the chevron folds have more sweeping efficiency compared to the asymmetric folds. Because of this, the total entrapment percentage for chevron folds is slightly more compared to the asymmetric folds. The distance and area covered by the CO_2 plume are high, so more traps were encountered by the CO_2 plume in the chevron folds integrated domains compared to asymmetrical folds (see Fig. 3a and b). When the anticline dome is integrated on the plain sloping domain, the sweeping distances increased even more, so for this reason, the structural and residual trapping is more in the anticline integrated domain than the plain sloping domain. This trend was observed irrespective of the types of folds integrated on synthetic domains. It is evident that both folds have shown a similar visual trends in the trapping, see Fig. 3, for all three synthetic domains. By comparing both the folds, the Chevron folds have a low movable plume percentage, which indicates a higher trapping percentage is recorded compared to asymmetrical folds; this observation is consistent for all three synthetic domains. The sweeping distance in the high perturbed sloping domain is low compared to the other two domains for both types of folds. So, for this reason, the residual trapping is lower compared to the low perturbed sloping domain. But the structural trapping is higher compared to the low perturbed sloping domain because, in the high perturbed sloping domain, the synthetic geological perturbation confines the CO_2 plume movement. Therefore, more amount of CO_2 will get trapped structurally and restrict CO_2 in exploring the migration traps for residual trapping. Compared to the other two domains, the high perturbed domain has a more significant total entrapment percentage and a low movable plume recorded. And it was also observed that recorded movable CO_2

plume quantity dominates in the plain sloping domain by far compared to the other two domains.

4 Conclusions

In this work, an attempt was made to analyse the influences of caprock topography on structural and residual trapping. The study has provided that the caprock surface integrated fold structure influences the CO_2 plume migration and sweeping efficiency, which further affects the entrapment percentage. The presence of an anticline dome has increased the sweeping efficiency and residual trapping percentage. For the synthetic domain having more perturbations, the structural trapping is dominant, but the residual trapping seen less. The CO_2 sweeping plays a crucial role in the residual trapping; it is observed that when the sweeping efficiency or area cover by the CO_2 plume is larger, the CO_2 trapped in the residual trapping is increasing. Future work includes the consideration of reactive transport by including geochemical reactions that deals the solubility trapping and mineral trapping mechanisms.

Acknowledgement The authors would like to acknowledge Science and Engineering Research Board (SERB), India, for providing financial support under the Core Research Grant with file no. EMR/2017/02450.

References

1. Zhang D, Song J (2013) Mechanisms for geological carbon sequestration. In: Procedia international union of theoretical and applied mechanics IUTAM. Elsevier, pp 319–327. https://doi.org/10.1016/j.piutam.2014.01.027
2. Viebahn P, Vallentin D, Höller S (2014) Prospects of carbon capture and storage (CCS) in India's power sector—an integrated assessment. Appl Energy 117:62–75. https://doi.org/10.1016/J.APENERGY.2013.11.054
3. Niu B, Al-Menhali A, Krevor SC (2015) The impact of reservoir conditions on the residual trapping of carbon dioxide in Berea sandstone. Water Resour Res 51:2009–2029. https://doi.org/10.1002/2014WR016441
4. Nilsen HM, Syversveen AR, Lie KA et al (2012) Impact of top-surface morphology on CO_2 storage capacity. Int J Greenh Gas Control 11:221–235. https://doi.org/10.1016/j.ijggc.2012.08.012
5. Lie K-A, Nilsen HM, Andersen O, Møyner O (2016) A simulation workflow for large-scale CO_2 storage in the Norwegian North Sea. Comput Geosci 20:607–622. https://doi.org/10.1007/s10596-015-9487-6
6. Punnam PR, Krishnamurthy B, Surasani VK (2021) Investigations of structural and residual trapping phenomena during CO_2 Sequestration in Deccan Volcanic Province of the Saurashtra Region, Gujarat. Int J Chem Eng 2021:1–16. https://doi.org/10.1155/2021/7762127
7. Bachu S (2008) CO_2 storage in geological media: role, means, status and barriers to deployment. Prog Energy Combust Sci 34:254–273. https://doi.org/10.1016/j.pecs.2007.10.001

8. Hammond GE, Laboratories SN, Lichtner P (2012) PFLOTRAN: reactive flow & transport code for use on laptops to leadership-class supercomputers. Groundw React Transp Model 141–159. https://doi.org/10.2174/97816080530631120101 0141
9. Lie K-A (2019) An introduction to reservoir simulation using MATLAB/GNU octave. Cambridge University Press
10. Møll Nilsen H, Lie K-A, Andersen O (2015) Analysis of CO$_2$ trapping capacities and long-term migration for geological formations in the Norwegian North Sea using MRST-co2lab. Comput Geosci 79:15–26. https://doi.org/10.1016/j.cageo.2015.03.001
11. Prasanna Lakshmi KJ, Senthil Kumar P, Vijayakumar K et al (2014) Petrophysical properties of the Deccan basalts exposed in the Western Ghats escarpment around Mahabaleshwar and Koyna, India. J Asian Earth Sci 84:176–187. https://doi.org/10.1016/j.jseaes.2013.08.028

Isotopes for Improving Hydrologic Modeling and Simulation of Watershed Processes

Joe Magner, Brajeswar Das, Rallapalli Srinivas, Anupam Singhal, and Anu Sharma

Abstract Lake water quality management is an important component of watershed protection and restoration. Hydraulic residence time (HRT) is a guiding parameter for quantifying the extent of exposure of lake ecosystem to chemicals. The impact on the δD and $\delta 18O$-year amplitude depends on the lake shape, basin size, connectivity, geology and climate characteristics, perineal fluctuations in lake water mass balance, and hydraulic residence time. HRT in the lake can be dynamic. Usually, HRT is calculated based on the balance between lake's volume and its input and output parameters, but there is a relatively simple and approximate method that uses stable water isotopes to assess the direction and possible speed of runoff in a lake. This study establishes the importance of stable isotopes in hydrologic modeling.

Keywords Isotope fractionation · Hydraulic residence time · Lake management · Hydrology · Water quality

1 Introduction

Isotope are elements having equal number of protons but unequal number of neutrons. The additional number of neutron or the absence of a neutron does not alter any of its chemical properties. But a relatively small variation in the mass of individual atoms expresses itself when a medium undergoes phase transition (from a vapor phase to a liquid phase). The compounds containing the heavier isotope atoms form heavier compound require relatively more specific energy for transition from liquid to vapor

J. Magner (✉)
Department of Bioproducts and Biosystems Engineering, University of Minnesota, Minneapolis, USA
e-mail: jmagner@umn.edu

B. Das · R. Srinivas · A. Singhal
Deparment of Civil Engineering, Birla Institute of Technology and Science, Pilani, India

A. Sharma
Department of Chemistry, University College of Engineering & Technology, Bikaner, India

© The Author(s), under exclusive license to Springer Nature Singapore Pte Ltd. 2022   123
R. Srinivas et al. (eds.), *Advances in Computational Modeling and Simulation*,
Lecture Notes in Mechanical Engineering,
https://doi.org/10.1007/978-981-16-7857-8_10

phase in case of evaporation as compared to condensation. This gives rise to isotopic fractionation, which a modification of the isotopic composition of water caused by the change from one phase to another phase.

Water is composed of hydrogen and oxygen. There are two stable isotopes of hydrogen:1H protium the lighter isotope and D (deuterium), twice as heavier. The third isotope, 3H or tritium, has radioactive property having a half-life of 12.3 years. 99.985% of the hydrogen atoms in the hydrosphere are of 1H isotope, and 2H only accounts for 0.015% [1]. Oxygen has three stable isotopes: 99.859% of 16O in the hydrosphere, 0.038% of 17O, and 0.2% of 18O. Isotopes account for a very small part of the total, but this difference can be measured, determined, and expressed as the ratio of the lightest to heaviest isotope [2, 3]. One can infer about watershed characteristics based on variation of the number of different isotopes. A standard called VSMOW (Vienna Standard Ocean Water Index) agreed with the IAEA (International Atomic Energy Agency) is used to determine the difference between the ratio of heavy isotopes and light isotopes [4].

During a storm, molecules with heavier isotopes are more likely to condense into sediments than molecules with lighter isotopes. As water moves in the hydrological cycle, the evaporation and condensation process redistribute the relative abundances of δD and $\delta 18O$. Values of $\delta 2H$ and corresponding $\delta 18O$ in precipitation are interdependent and follow a linear relationship on a global scale [5]. This relationship is defined as 'Global Meteoric Water Line' (GMWL) with a slope of 8 and an intercept of 10 (Fig. 1).

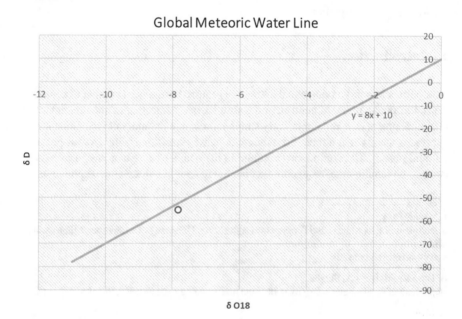

Fig. 1 Global meteoric water line [6, 7]

Rozansky et al. [8] determined Eq. (1) based on data from 206 global precipitation collection stations in the IAEA network:

$$\delta^2 H = 8.17 \pm 0.06 \cdot \delta^{18}O + 10.35 \pm 0.65\% \tag{1}$$

Kumar et al. [9] reported the Meteorological Water Level (MWL) in India and several regional waterways (RMWL) in northern India, southern India, and western Himalayas (Eqs. 2–5). The differences in the slope and intersection of these lines are due to different geographic and meteorological condition.

- Meteoric Water Line for India (MWL_I):

$$\delta^2 H = 7.93 \pm 0.06 \cdot \delta^{18}O + 9.94 \pm 0.51\% \left(n = 272; r^2 = 0.98\right) \tag{2}$$

- Regional Meteoric Water Line for northern India:

$$\delta^2 H = 8.15 \pm 0.12 \cdot \delta^{18}O + 9.55 \pm 0.80\%0 \left(n = 65; r^2 = 0.99\right) \tag{3}$$

- Regional Meteoric Water Line for southern India:

$$\delta^2 H = 7.82 \pm 0.17 \cdot \delta^{18}O + 10.23 \pm 0.85\%0 \left(n = 62; r^2 = 0.97\right) \tag{4}$$

- Regional Meteoric Water Line for thewestern Himalayas:

$$\delta^2 H = 7.95 \pm 0.09 \cdot \delta^{18}O + 11.51 \pm 0.89\% \left(n = 123; r^2 = 0.99\right) \tag{5}$$

2 Calculation of Hydraulic Residence Time (HRT)

Lake's ecosystem equilibrium is greatly dependent on water quality indicators which in turn are governed by the HRT of the lake. Hydraulic residence time is a guiding parameter for quantifying the exposure of lake ecosystem to chemicals. For calculating HRT, many variable parameters of the lake like watershed size, volume, location within a watershed and climatic variability need to be considered. Owing to complex network of inputs and outputs and their interdependencies, calculation of exact HRT becomes extraneous.

To quantify the water balance of the lake and assess its sensitivity to pollutant penetration, the increased residence time in the water is a decisive parameter. Higher HRT would imply prolonged exposure to chemical agents which in turn would result in growth of cyano bacteria [10] There are several techniques for calculating HRT including but not limited to Lacustrine water budgets, groundwater flow net analysis, hydro-chemical lake water budgets, tracer analysis, thermal indicators, seepage parameter, and biological parameters [11, 12]. The difficulty with these methods is that the result depends on small pressure drops and small-scale changes in hydraulic conductivity would be not be measured. Estimates based on stable isotopes are highly dependent on relative humidity and the isotopic composition of atmospheric vapor and evaporation [13]. All these parameters are error-prone; however, the use of stable isotopes of water requires relatively less data points and the uncertainty in the estimating hydraulic residence time is less. The works demonstrate the ability of stable isotope technology to estimate the long-term average residence time in lake water recharged by groundwater based on a relatively small amount of data. Lake management variable data and groundwater isotope composition, accompanied by complete and high-quality long-term meteorological and isotopic data (precipitation) from nearby measuring stations, can be used to simulate the annual cycle of the stable water isotope inventory in the lake, which can then be compared with the lake inventory observation value at a specific time.

2.1 Calculation of HRT Using Isotope Mass Balance

The technical interpretation of hydraulic radius varies for different stakeholders, some use water discharge rates, and others use groundwater outflow rate and evaporation. The use of $\delta18O$ and $\delta2H$ is to determine Lacustrine groundwater discharge (LGD), and lake residence time is based on isotope mass balance. This requires the isotope composition of all components of the lake water balance, where δL, δP, δGi, δGo, and δE are the isotope composition of lakes, precipitation, groundwater runoff, seepage water, and evaporation. The annual isotope mass balance under the assumption of constant lake volume over time is given using Eq. (6).

$$P\delta_p + G_i\delta_{Gi} = E\delta_E + G_o\delta_{G_0} \tag{6}$$

The dynamic isotope mass balance for a well-mixed lake can be written using Eq. (7):

$$\delta_{L_{t+1}} = \delta_{L_t} + \frac{\left[P_t\delta_{P_t} + G_i\delta_G - E_t\delta_{E_t} - G_0\delta_{G_0}\right]}{V} \tag{7}$$

The equation can be rearranged and solved for Gi or Go. Although δL, δP, and δGi can be measured directly, it is usually assumed that δGo is equal to δL. δE is not easy to measure. Since evaporation is the process that determines the evolution of the isotope composition of lakes, its accurate estimation is critical to the accuracy of water balance. δE is calculated using the linear resistivity model of Craig and Gordon [6] (Eq. 8), which describes δE as a function of relative humidity, temperature, lake surface isotope composition (δLs), and air humidity (δA isotope composition).

$$\delta_E = \frac{\frac{\delta_s - \varepsilon^+}{a^+} - h\delta_A - \varepsilon_K}{1 - h + 10^{-3}\varepsilon_K} \tag{8}$$

The variables in Eq. (8) are the equilibrium isotopic separation ε + (temperature dependent), the equilibrium isotopic fraction factor α + (temperature dependent), the kinetic isotopic separation εK (humidity dependent), and the relative humidity h [−]. It is possible to measure or estimate δA from δP and air temperature. Compensation is necessary because δA and δP are usually out of balance throughout the year under seasonal weather conditions. The seasonal coefficient k ranges from 0.5 for strong seasonal climates to 1 for non-seasonal climates and is estimated by dual analysis of δ2H and δ18O. ΔA weighted by the annual average of the evaporation flow is corrected (optimization of k) to adapt δE (Eq. 9) to the local evaporation line.

$$\delta_A = \frac{\delta p - k\varepsilon^+}{1 + 10^{-3}k\varepsilon^+} \tag{9}$$

3 Case Study: Minnesota's Sentinel Lakes

Amplitude of fractionation of lake water and water vapor was studied to predict the hydraulic retention time of 24 lakes in Minnesota. Figure 2 shows local meteoric water line of Minnesota sentinel lakes. Over a period of three years, the lakes were

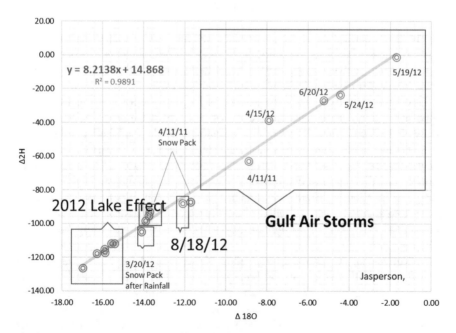

Fig. 2 Local meteoric water line (LMWL) of Minnesota sentinel lakes

sampled in spring, summer, and autumn. The results indicate that the annual contribution of water sources, basin area and connectivity have all changed. The calculated annual interval of HRT is 18.8 to 0.4 years. The annual amplitudes of δD and δ18O provide directional information about the residence time of each lake and record seasonal changes in the components of lake D and δ18O. δD and δ18O can provide water quality managers with a cost-effective tool to better understand, protect, and restore lakes and their catchment areas.

From 2008 to 2010, 24 lakes were visited in May, July, and October respectively, and water samples were collected for δD and δ18O stable isotope analysis. The sampling period is kept exactly the same as the crop rotation in spring and autumn and midsummer. In the case of maximum evaporation, sampling during these important time periods is the best way to capture the δD and δ18O fluctuations of each lake. Comparison of the composition of stable isotopes with the isotopic composition of water vapor in the atmosphere, which has known that isotopic concentrations at certain latitudes and temperatures have been performed. In order to predict the hydraulic retention time δ18O, the deviation between the fractionation amplitude and seawater is modeled by water vapor. The estimated value of the seasonal value of water vapor is determined based on the minimum and maximum average seasonal temperature using Eq. (10)

$$\delta^{18}O = (0.521 \pm 0.014) \times T(C) - (14.96 \pm 0.21) \tag{10}$$

The minimum and maximum seasonal temperatures are calculated to represent the expected range of stable atmospheric water vapor isotopic composition for each lake. The maximum seasonal range is modeled with the isotopic concentration observed in the lake water to estimate the hydraulic retention time.

By comparing the width of the best-fit curve for precipitation with the width of the same curve for the water of interest, equation (original HRT) is used to estimate the hydraulic residence time. The seasonal variation of δ18O precipitation components in temperate regions tends to be sinusoidal. This pattern can be observed throughout the year, reflecting seasonal changes in tropospheric temperature. The measurement changes of δ18O components in rivers, lakes, ponds, groundwater, or groundwater at a given location in different seasons has been obtained. When seasonal water is considered stationary and well mixed with the exponential distribution of residence time, the average value of the hydraulic system can be calculated using the original HRT equation.

In Minnesota, the D and δ18O trends of the four Major Land Use Regions (MLRUs) studied were similar. In Lake Minnesota, there was a general transition from lighter to higher values for δD and δ18O values from north to south. This was the result of weather conditions unique to the Canadian MLUR shield. The weather system in this area is usually originated in the Arctic, and compared with the weather system originated in the Gulf of Mexico, it produced a small amount of isotopic water vapor source. The lake in the study and the Minnesota evaporation line showed a

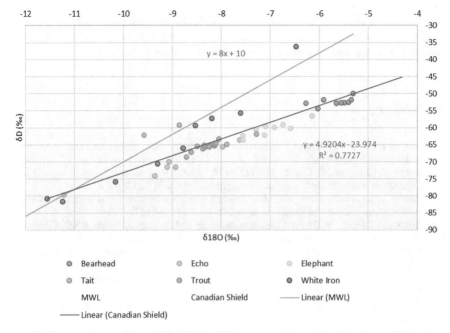

Fig. 3 Seasonal Isotope pattern (Canadian shield)

strong correlation with $R^2 = 0.950$ (Fig. 3). The deviation from MWL could be seen on various scales. The results are interpreted by state, Major Land Use Regions (MLUR), and individual lakes. The evaporation line shows that the lake water is richer in $\delta 18O$. Compared with the other three MLURs, the Canadian Shield Lake tracked below and near the MWL is a water body with very low oxygen content.

4 Conclusions

Based on the deviation of ratios stable isotopes δD and $\delta 18O$, annual and seasonal variations of water source and hydraulic residence time can be identified, taking into account the potential of pollutant input in the catchment area. If best management practices (BMP) are not followed, it is likely to release large amounts of pollutants into the lake. The D and $\delta 18O$ amplitudes of some lakes are very conservative. The basin and lake model uses hydraulic retention time to determine amount of time of exposure of pollutants to the lake. Usually, the hydraulic residence time is calculated based on the balance between the volume of the lake and the lake's input and output. Compared with physical mass balance methods, δD and $\delta 18O$ data can be used to understand hydraulic retention time and water supply contribution faster and cheaper. The determination of the proportion of water sources and the influence of seasonal fluctuations on the composition of D and $\delta 18O$ are of decisive significance for the determination of lake pollution. Analysis and decision-making considering isotopic composition data can revolutionize lacustrine watershed management systems and improve targeted and optimized catchment management.

References

1. Magner JA, Regan CP, Trojan M (2001) Isotopic characterization of the buffalo river watershed and the buffalo aquifer near Moorhead municipal well one. Hydrol Sci Technol 17:237–245
2. Asano Y, Compton JE, Church MR (2006) Hydrologic flow paths influence inorganic and organic nutrient leaching in a forest soil. Biogeochemistry 81(2):191–204
3. Asano Y, Uchida T, Ohte N (2002) Residence times and flow paths of water in steep unchannelled catchments, Tanakami Japan. J Hydrol 261(1–4):173–192
4. Minet EP, Goodhue R, Meier-Augenstein W, Kalin RM, Fenton O, Richards KG, Coxon CE (2017) Combining stable isotopes with contamination indicators: a method for improved investigation of nitrate sources and dynamics in aquifers with mixed nitrogen inputs. Water Res 124:85–96
5. Simpson HJ, Herczeg AL (1991) Stable Isotopes as an indicator of evaporation in the river Murray. Water Resour Res 27:1925–1935
6. Craig H, Gordon LI (1965) Deuterium and oxygen 18 variations in the ocean and the marine atmosphere
7. Jasperson JL, Gran KB, Magner JA (2018) Seasonal and flood-induced variations in groundwater-surface water exchange in a northern coldwater fishery. J Am Water Resour Assoc 54(5):1109–1126. https://doi.org/10.1111/1752-1688.12674
8. Rozanski K, Araguás-Araguás L, Gonfiantini R (1993) Isotopic patterns in modern global precipitation. Geophys Monogr-Am Geophys Union 78:1–1
9. Kumar B, Rai SP, Kumar US, Verma SK, Garg P, Kumar SV, Pande NG (2010) Isotopic characteristics of Indian precipitation. Water Resour Res 46(12)
10. Magner J, Zhang L (2014) Cross river watershed hydrologic adjustment pre-& post-June 2012 mega-storm. J Environ Sci Eng B 3:133–141
11. Petermann E, Gibson JJ, Knöller K, Pannier T, Weiß H, Schubert M (2018) Determination of groundwater discharge rates and water residence time of groundwater-fed lakes by stable isotopes of water (18O, 2H) and radon (222Rn) mass balances. Hydrol Process 32(6):805–816
12. Engel L, Magner J (2019) Estimating the direction of lake hydraulic residence in Minnesota's sentinel lakes: implications for management. Int J Hydro 3(5):362–367
13. Romo S, Soria J, Fernandez F, Ouahid Y, Baron-Sola A (2013) Water residence time and the dynamics of toxic cyanobacteria. Freshw Biol 58:513–522

Kinetic Energy Correction Factor for a Converging–Diverging Nozzle

Sadhya Gulati, Snehaunshu Chowdhury, and Eldhose Iype

Abstract Kinetic energy correction factor is often neglected while using Bernoulli's equation for many applications. Here, flow through a converging–diverging (CD) nozzle is simulated to show that the kinetic energy correction factor (α) can take values between 1 and 3. A large reduction is observed in total mechanical energy in the diverging section of the nozzle when α is not accounted for. A negative correlation is observed between Reynolds number and α.

1 Introduction

Bernoulli's equation explains the relationship between pressure and fluid velocity for inviscid fluid flows. In its simplest form, the equation is an expression of mechanical energy conservation. For incompressible flows, this mathematical expression is written as

$$p + \frac{1}{2}\rho v^2 + \rho g h = constant, \tag{1}$$

where ρ is the density, p is the pressure, v is the velocity, h is elevation measured with respect to any arbitrary datum, and g ($= 9.8\,\mathrm{ms}^{-2}$) is the gravitational acceleration directed vertically downward. This implies that the sum of kinetic energy, gravitational potential energy, and pressure energy remains constant along a streamline. One of the assumptions in using Bernoulli's equation is that the flow is frictionless.

S. Gulati · E. Iype (✉)
Department of Chemical Engineering, BITS Pilani, Dubai Campus, Dubai, UAE
e-mail: e.iype@dubai.bits-pilani.ac.in

S. Gulati
e-mail: f20160191d@alumni.bits-pilani.ac.in

S. Chowdhury
Department of Mechanical Engineering, BITS Pilani, Dubai Campus, Dubai, UAE
e-mail: snehaunshu@dubai.bits-pilani.ac.in

© The Author(s), under exclusive license to Springer Nature Singapore Pte Ltd. 2022   131
R. Srinivas et al. (eds.), *Advances in Computational Modeling and Simulation*,
Lecture Notes in Mechanical Engineering,
https://doi.org/10.1007/978-981-16-7857-8_11

Meaning, the viscous effects are negligible. But in practical situations (especially in internal flows), this assumption is often violated. Prior experiments show that total mechanical energy is not constant across the cross section of flow [2, 3]. In order to account for this variation, kinetic energy correction factor (α), i.e. the ratio between actual kinetic energy and the kinetic energy with uniform velocity profile, is introduced in the Bernoulli's equation. This is defined as below.

$$\alpha = \frac{1}{A} \int (u/v)^3 dA. \tag{2}$$

The modified Bernoulli's equation becomes

$$p + \alpha \frac{1}{2} \rho v^2 + \rho gh = constant. \tag{3}$$

Here, A is the cross-sectional area, u is the velocity at any point within the cross section, and v is the average velocity. The kinetic energy correction factor considers that the velocity profile is not uniform and estimates the difference between the actual velocity distribution and an idealized uniform profile, viz., the average velocity. For a fully developed laminar flow through a pipe, α value is around 2, whereas for turbulent flow, it is about 1.05 [3]. The average of α values obtained for a compound channel flume is reported to be 1.094, for regular channels is 1.15 [4], for natural streams and torrents, it is 1.30, and for river under ice cover is 1.50 [5]. Fenton [6] shows that α values for a turbulent logarithmic velocity distribution in moderately rough pipes and channels range from 1.05 to 1.1, whereas for a small irrigation channel it ranges from 1.10 to 1.20 [7]. The literature, therefore, shows a significant variation of α values depending on the flow geometries and conditions.

A commonly used geometry in many practical applications is the converging–diverging (CD) nozzle. Such geometries are used for supersonic applications [8], flashing flows [9], venturi meter, etc., to name a few. To the best of our knowledge, α values have never been reported for a CD nozzle. CD nozzle is important because most of the experiments used to validate Bernoulli's equation are performed on a CD nozzle. Some experiments on the flow through sudden contractions and expansions were performed by McNeil and Morris [10]. However, the α values for such geometries are missing. Sudden expansion creates more flow eddies than sudden contraction, and therefore more losses occur at the point where expansion in the pipe begins [11], i.e., the diverging section of the CD nozzle. For an internal flow undergoing sudden expansion, kinetic energy is lost and pressure is not fully recovered [12]. A static pressure change, which results in energy loss, is observed across the expansion [13] cross section. In this present study, we have simulated the flow through a CD nozzle, and numerically investigated the changes in kinetic energy correction factor (α) for an incompressible Newtonian fluid (water) at different flow rates including both laminar and turbulent regions. This is expected to fill the void in literature on (α) for flows through CD nozzles.

2 Methodology

A two-dimensional axi-symmetric simulation of flow through a CD nozzle is conducted using ANSYS Fluent software. Water is chosen as the fluid flowing through the CD nozzle. Default values in Fluent are used for relevant properties of water. A 60 cm long converging–diverging nozzle (equal length on both converging and diverging sections) with a throat in the middle (Fig. 1) is used to simulate the flow. The inner radii at the entrance and exit are 3 cm each, and that of the throat is 1.5 cm. In order to ensure fully developed flow near the entrance of the nozzle and to eliminate end-effects, the pipe is extended by 13.5 cm in both the directions. The geometry and the mesh are shown in Fig. 1. Special care is taken to have finer mesh adjacent to the walls to capture viscous gradients as this is important in calculation of α.

The volume flow rates are chosen to represent flows in the laminar, transition, and turbulent regimes for the given geometry based on inlet dimensions. The details are given in Table 1.

The outlet is kept as a velocity outlet with the same velocity as the inlet for all flow rates. The axis of the CD nozzle is a symmetry line, while the outer radii at any axial location is defined as a solid wall. The flow is modeled using k-epsilon model. After the solution is converged, velocity and pressure values at nine equidistant cross-sectional surfaces between the entry and exit of the CD nozzle for each flow rate are noted. The surface integrals of velocities were computed for all the cross-sectional areas in order to calculate the kinetic energy correction factor, α. A grid convergence test confirmed that further mesh refinement was not necessary. The final optimized mesh contained 11,200 cells and 11,781 nodes with a maximum aspect ratio of 11.6. There were no red flags or convergence issues raised by the solver.

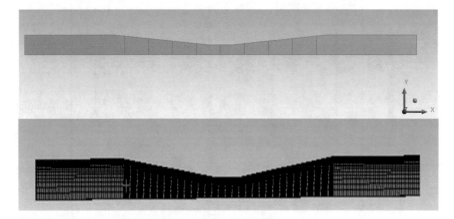

Fig. 1 Geometry and mesh of the converging–diverging nozzle

Table 1 Flow rates and inlet velocities used for simulation

Regime	Volume flow rate (LPH)	Inlet velocities (m/s)
Laminar and transition	120	0.0541
	160	0.0721
	200	0.0902
	240	0.108
Turbulent	800	0.361
	900	0.406
	950	0.428
	1000	0.451

3   Results

3.1   Kinetic Energy Correction Factor for Various Flow Rates

Different values of kinetic energy correction factors were obtained using Eq. (2), ranging from 1 to 3.13, for all 9 locations at different flow rates as shown in Fig. 2. For the converging section, the α values were found to be almost constant for all the flow rates; however, the alpha values were found to be decreasing with increasing flow rate for diverging section (see inset in Fig. 2). For flows toward turbulent regime, the local velocity (u) becomes more and more uniform across the cross section, and

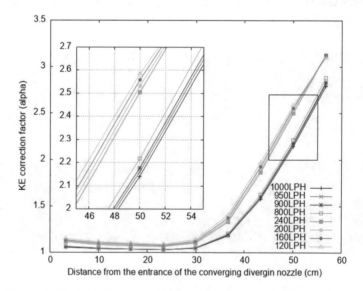

Fig. 2 Changes in kinetic energy correction factor (α) at various location in the converging–diverging nozzle for various flow rates

therefore, the local velocity becomes closer to the average velocity (v). This causes the value of α to decrease. It has been reported in the literature that as the flow rate increases, α values decrease and eventually becomes constant [5]. However, the influence of viscous effects is known to increase the value of α [14]. At higher Reynolds number (turbulent regime), contribution of viscous losses is lower than kinetic energy factor due to cross sectional changes. Therefore, the loss of kinetic energy is taken into account and frictional effects (viscosity factor) are neglected [15]. Thus, the accuracy of the Bernoulli equation is improved for higher Reynolds number as the kinetic energy term dominates and hence lower α values are required for more turbulent flow [16].

3.2 Total Mechanical Energy Conservation with Kinetic Energy Correction

The total mechanical energy, i.e. the sum of pressure energy, kinetic energy, and gravitational potential energy, was calculated at all the 9 cross-sections for different flow rates. The changes in the total energy across all nine positions are shown in Fig. 3a for turbulent flow and in Fig. 3b for laminar or transition flow, respectively. The kinetic energy is multiplied by the kinetic energy correction factors (α) obtained above and the corrected total energy profiles are also given. For turbulent flow, the energy is almost constant near the entrance; however, it decreases as the fluid moves along the nozzle. This is because of large frictional losses at the diverging section. However, when kinetic energy correction factor is added, the rate of decrease in total energy reduces. In the case of laminar and transition regimes (Fig. 3b), the frictional losses are almost negligible, and therefore there was barely any change in total energy while using kinetic energy correction factor.

3.3 Kinetic Energy Correction Factor verus Reynolds Number

The diameter of the CD nozzle changes with axial position. Obeying the continuity relation, this implies a change in the velocity of the liquid axially. This also leads to a variation of local Reynolds number as both the diameter and the velocity changes with position. Thus, an analysis of the variation of kinetic energy correction factor with Reynolds number gives an insight into a potential correlation for this geometry. Figure 4a shows the variation of kinetic energy correction factor with Reynolds number for the converging section. As can be seen, the kinetic energy correction factor decreases with Reynolds number almost hyperbolically. At very high Reynolds numbers, α almost reaches a value of unity, which corresponds to the turbulent flow correction factor for flow through a pipe. In the case of diverging section of the pipe,

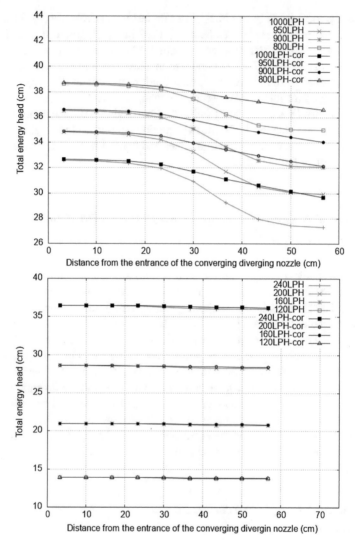

Fig. 3 Total energy (sum of potential, kinetic and pressure energy) with and without kinetic energy correction as a function of position along the nozzle. **a** turbulent flow **b** laminar or transition flow

the magnitude of α is significantly higher than that for converging section (compare the y-axis values for the same). Although α seems to be slightly correlated with Reynolds number for the diverging section, the values are mostly dependent on the diameter of the location. This is evident from the clusters that we see corresponding to individual locations within the nozzle that we see in Fig. 4b.

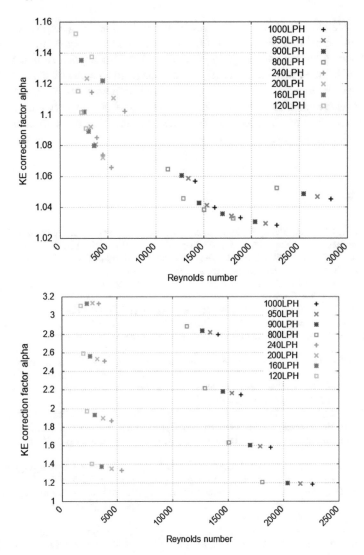

Fig. 4 Kinetic energy correction factor versus Reynolds number for **a** Converging section, **b** diverging section of the nozzle

4 Conclusion

In this study, we analyzed the kinetic energy correction factor (α) in the Bernoulli equation for a converging–diverging nozzle at different flow rates representing laminar, transition, and turbulent regimes. For the converging section, the α values are almost close to unity while for the diverging section, it can go up to 3. This is attributed to large pressure drag or form drag due to boundary layer separation and recirculation in the diverging section of the nozzle. For the diverging section, the laminar and transitional flow show different characteristics compared to the turbulent flow. Whereas, for the converging portion, they show gradual and similar behavior. It is found that the use of kinetic energy correction factor is essential in the diverging section in order to get a correct energy profile for turbulent flows. We also conclude that an inverse relationship exists between α and Reynolds number in the converging part, whereas for the diverging part, α values were found to be dependent only on their location in the nozzle.

References

1. Fester V, Mbiya B, Slatter P (2008) Energy losses of non-Newtonian fluids in sudden pipe contractions. Chem Eng J 145(1):57–63. https://doi.org/10.1016/j.cej.2008.03.003
2. Saleta ME, Tobia D, Gil S (2005) Experimental study of Bernoulli's equation with losses. Am J Phys 73(7):598–602. https://doi.org/10.1119/1.1858486
3. McCabe WL, Smith JC, Harriott P (1993) Unit operations of chemical engineering, vol 5. McGraw-hill New York
4. Te Chow V (1958) Open channel hydraulics. McGraw Hill Publications
5. Seckin G, Ardicilioglu M, Cagatay H, Cobaner M, Yurtal R (2009) Experimental investigation of kinetic energy and momentum correction coefficients in open channels. Sci Res Essay 4(5):473–478
6. Fenton JD (2005) On the energy and momentum principles in hydraulics. In: Proceedings of the Korea water resources association conference. Korea Water Resource Association
7. Garba Wali U (2013) Kinetic energy and momentum correction coefficients for a small irrigation channel. Int J Emerg Technol Adv Eng 3(9):8
8. Park J-J et al (2011) Supersonic nozzle flow simulations for particle coating applications: effects of shockwaves, nozzle geometry, ambient pressure, and substrate location upon flow characteristics. J Therm Spray Technol 20(3):514–522. https://doi.org/10.1007/s11666-010-9542-8
9. Liao Y, Lucas D (2015) 3D CFD simulation of flashing flows in a converging-diverging nozzle. Nucl Eng Des 292:149–163. https://doi.org/10.1016/j.nucengdes.2015.06.015
10. McNeil DA, Morris SD (1996) A mechanistic investigation of laminar flow through an abrupt enlargement and a nozzle and its application to other pipe fittings. Publications Office of the EU. Report EUR 16348. Dated: 24 Mar 1996
11. Satish G, Ashok Kumar K, Prasad VV, Pasha SM (2013) Comparison of flow analysis of a sudden and gradual change of pipe diameter using FLUENT software. Int J Res Eng Technol 2(12):41–45
12. Rijsterborgh H, Roelandt J (1987) Doppler assessment of aortic stenosis: bernoulli revisited. Ultrasound Med Biol 13(5):241–248. https://doi.org/10.1016/0301-5629(87)90096-2
13. Lottes PA (1961) Expansion losses in two-phase flow. Nucl Sci Eng 9(1):26–31. https://doi.org/10.13182/nse61-a25860
14. Cassan L, Belaud G (2010) Experimental and numerical studies of the flow structure generated by a submerged sluice gate. In: Proceedings from first IAHR European congress. Edinburgh, Scotland, 64-6 May 2010
15. Wibowo WA, Pranolo SH, Sunarno JN, Purwadi D (2014) Effect of biomass feed size and air flow rate on the pressure drop of gasification reactor. J Teknol (Sci Eng) 68(3):13–16. https://doi.org/10.11113/jt.v68.2962
16. Etemad SG, Thibault J, Hashemabadi SH (2003) Calculation of the Pitot tube correction factor for Newtonian and non-Newtonian fluids. ISA Trans 42(4):505–512. https://doi.org/10.1016/s0019-0578(07)60001-9

Modeling of Fluid-Structure Interactions with Exact Interface Tracking Methods

Pardha S. Gurugubelli and Vaibhav Joshi

Abstract Accuracy and numerical stability of nonlinear coupled fluid-elastic inter-action simulations largely depends on the coupling and interface modeling algo-rithms. As part of the numerical coupling, the coupled solver requires to satisfy the kinematic and dynamic continuity conditions along the interface in addition to the fluid and structural dynamics governing equations. The interface kinematics and dynamics conditions are traditionally coupled with the governing equations that define the dynamics using either a partitioned or monolithic approaches. Irrespective of the coupling approach considered, the accuracy of the coupled numerical simula-tions strongly depends on the accuracy of the structural response dynamics, which in turn depends on the accuracy of the fluid dynamic forces acting on the struc-ture. Hence, the numerical methods with exact interface become attractive when the accuracy of the coupled fluid-elastic interactions is of importance. In this paper, we present a review on the class of quasi-monolithic approaches with exact interface for numerically modeling the fluid-structure interactions involving rigid and flexible multi-body systems.

1 Introduction

Fluid-Structure Interaction (FSI) is a branch of multi-physics that is commonly observed in our day-to-day life, wherein the structure is considered to be elastic and it can undergo deformation/displacement due to the fluid dynamic forces acting on it which in turn would influence the fluid dynamic forces acting on them [1, 2]. As

P. S. Gurugubelli (✉)
Department of Mechanical Engineering, Birla Institute of Technology & Science Pilani,
Hyderabad Campus, Hyderabad, India
e-mail: pardhasg@hyderabad.bits-pilani.ac.in

V. Joshi
Department of Mechanical Engineering, Birla Institute of Technology & Science Pilani,
K K Birla Goa Campus, Goa, India
e-mail: vaibhavj@goa.bits-pilani.ac.in

© The Author(s), under exclusive license to Springer Nature Singapore Pte Ltd. 2022   139
R. Srinivas et al. (eds.), *Advances in Computational Modeling and Simulation*,
Lecture Notes in Mechanical Engineering,
https://doi.org/10.1007/978-981-16-7857-8_12

a result, the aerodynamic/hydrodynamic forces acting on an elastic structure would be distinctly different from that of a fixed structures [3]. Hence, it is important to investigate the FSI in the design of tall buildings, buildings housing sensitive equipment, long bridges [4], off-shore floating platforms [5], oil and gas pipelines [6, 7], wind turbines, micro-air vehicles, etc.

Numerical modeling of coupled FSI would require satisfying the interface kinematic (i.e., velocity and displacements) and dynamic (i.e., force) continuity along the fluid-structure interface along with to the fluid and structural dynamic governing equations. A FSI computational models can be categorized based on the way the interface between fluid-structure is modeled and the way in which the interface conditions are satisfied along the interface. The interface between the fluid-structure is either modeled by a conforming/non-confirming body fitted mesh for the fluid and structural domains or by using a non-body fitted Cartesian grid mesh for the combined fluid-structure domain. Methods such as level set method [8], Lagrangian multiplier, immersed boundary [9], ghost fluid, and fictitious domain [10] methods come under the later category of the interface modeling methods. On the other hand, arbitrary Lagrangian-Eulerian (ALE) is one of the popular approaches that would come under the body fitted mesh category of the interface modeling. In this coordinate system, the computational nodes can move relative to the spatial coordinate system. In an ALE approach, the mesh nodes on the fluid-structure interface behave like material points in a Lagrangian frame and the nodes inside the fluid domain can be moved arbitrarily to account for the movement of the fluid-structure interfaces so that the fluid and the structural meshes always remain two distinct non-overlapping meshes [11]. Hence, a typical body fitted FSI is a three-field problem [12].

Traditionally, in an ALE-based approach, the interface conditions are satisfied by a partitioned or monolithic approach. In a partitioned approach, the structural dynamic, the interface kinematic, dynamic continuity boundary conditions, and the fluid dynamic governing equations are solved in a sequential order [13]. As the governing equations for each of the physical fields, i.e., the fluid and structure, are solved sequentially one can use existing traditional fluid and the structural solvers as black-box solver by transferring the fluid forces acting on the solid from the fluid-solver to the structural-solver and transferring the structural kinematics, i.e., displacement and velocity, from the structural-solver to the fluid-solver. Since any fluid/structural solvers can be coupled with each other to simulate the coupled FSI phenomena, the partitioned approach offers higher levels of flexibility and modularity. Due to these traits, partitioned-based approaches are popular. However, as the effects of fluid and elastic structures on each other are transferred as boundary conditions coupling between the fluid and solid is not strong enough and can lead to numerical instability when the structural mass of the structure interacting with surrounding fluid is of the same order or lower because of the spurious energy produced due to the temporal inaccuracies [14]. This numerical instability can be solved for certain low mass cases by satisfying the interface conditions over multiple iterations till the solutions achieve a kinematic and dynamic convergence [15]. Even this strong fluid-structure coupling over multiple iterations may not be enough to sustain a numerical stability for very low mass structures and the solutions can either suffer from non-convergence or convergence to a wrong solution-related issues.

In a monolithic approach, the governing equations that define the fluid dynamics, structural dynamics, interface continuity, and mesh dynamics are all assembled into a single large matrix and solved together [16]. These schemes provide good numerical stability even for problems with very low mass structures that experience strong inertial effects. However, monolithic approaches lack the flexibility and modularity of using an established fluid/structural solvers. In addition to the lack of flexibility, monolithic approaches can suffer from the computational resource- and convergence-related issues for solving large ill-conditioned system of linear equations. This would necessitate development of special kind of pre-conditioners for solving the matrix system of equations.

Key objective of the current work is to present a review on ALE-based FSI formulations which are numerically stable and computationally efficient for low structure to fluid-mass ratio where the inertial effects are very strong. As part of the review, we have considered two FSI coupling formulations based on improvised monolithic approach [17–19]. Unlike the traditional monolithic approaches wherein the governing equations pertaining to fluid, structure, interface, and mesh update are all solved together in a single large matrix, in the improvised monolithic approaches the equations pertaining to the mesh update algorithm are decoupled from the fluid, structure, and interface continuity equations by explicitly predicting the structural positions at the start of each time step. Explicit prediction on structural positions also enables linearization of the convective terms. Additionally, in these schemes, the interface kinematic and dynamic continuity conditions are implicitly satisfied by the construction of a single unified governing equation for combined fluid-structure system.

In the current work, we begin with a brief overview of the governing equations involved in the numerical modeling in Sect. 2. In Sect. 3, we construct the variational weak form of the combined fluid-structure governing equations. We then present in the complete second-order time-accurate quasi-monolithic with exact interface tracking formulation in Sect. 4. We then present Galerkin least square stabilized quasi-monolithic formulation in Sect. 5. We conclude the work by providing a summary of the two quasi-monolithic solvers reviewed in this work in Sect. 6.

2 Governing Equations

Let us consider x as a spatial point that belongs to the three-dimensional fluid domain $\Omega^{f(t)}$ with boundary $\Gamma^f(t)$ at any time t and can move randomly in $\Omega^{f(t)}$. The domain boundary Γ^f is made up of the Dirichlet (Γ_D^f), Neumann (Γ_N^f), and the fluid-structure interface (Γ) boundaries. Similarly, let us also consider a material point Z corresponding to the initial three-dimensional structural domain Ω^s with boundary Γ^s that is made up of the Dirichlet (Γ_D^s), Neumann (Γ_N^s), and the interface (Γ). The

Navier–Stokes equations governing the dynamics of viscous incompressible fluid flow in an ALE reference frame are given as

$$\rho^f \frac{\partial u^f}{\partial t} + \rho^f \left(u^f - w \right) \cdot \nabla u^f = \nabla \cdot \left[-p I + \mu^f \frac{1}{2} \left(\nabla u^f + \left(\nabla u^f \right)^T \right) \right] + f^f \quad \text{in} \quad \Omega^{f(t)}, \quad (1)$$

$$\nabla \cdot u^f = 0 \quad \text{in} \quad \Omega^{f(t)}, \quad (2)$$

where ρ^f and μ^f are the fluid density and dynamic viscosity; $u^f = u^f(x, t)$, $w = w(x, t)$, and $p = p(x, t)$ represent the fluid velocity, fluid mesh velocity, and pressure defined at x for a time t; I is the second-order identity tensor; and f^f denotes the body force.

The structural dynamic of a flexible structure is typical governed by Navier's equation which is written as

$$\rho^s \frac{\partial^2 \eta^s}{\partial t^2} = \nabla \cdot \sigma^s + f^s \quad \text{in} \quad \Omega^s, \quad (3)$$

where ρ^s is the structural density and $\eta^s(Z, t)$ is the displacement vector that maps the material point Z from its initial position to its position at time t. σ^s denotes the first Piola-Kirchhoff stress tensor and f^s represents the body force vector acting on the structure. For a linear elastic material

$$\sigma^s \left(\eta^s \right) = \mu^s \left[\nabla \eta^s + (\nabla \eta^s)^T \right] + \lambda^s (\nabla \cdot \eta^s) I, \quad (4)$$

where μ^s and λ^s are the Lamé coefficients of a material satisfying [20]. Similarly, the constitutive relation for a St. Venant-Kirchhoff (SVK) material [20, 21] is

$$\sigma^s \left(\eta^s \right) = 2\mu^s F E + \lambda^s \left[\text{tr} \left(E \right) \right] F, \quad (5)$$

where $\text{tr}(\cdot)$ is the tensor trace operator, F and E represent the deformation gradient and the Green-Lagrangian strain tensors, respectively, and are given as [20]

$$F = (I + \nabla \eta^s), \quad E = \frac{1}{2} \left(F^T F - I \right). \quad (6)$$

For simplicity, one can rewrite the structural Eq. (3) as

$$\rho^s \frac{\partial u^s}{\partial t} = \nabla \cdot \sigma^s + f^s \quad \text{in} \quad \Omega^s, \quad (7)$$

considering

$$u^s(Z, t) = \frac{\partial \varphi^s}{\partial t}. \quad (8)$$

The above simplification enables the implicit implementation of the kinematic continuity condition along the fluid-structure interface.

The fluid and structural dynamics governing equations presented above need to satisfy the Dirichlet and Neumann conditions along the respective non-interface domain boundaries, which can be expressed below:

$$u(x, t)^f = u_D^f \quad \forall x \in \Gamma_D^f \quad \text{and} \quad \sigma^f(x, t) \cdot n^f = \sigma_N^f \quad \forall x \in \Gamma_N^f. \tag{9}$$

$$u^s(Z, t) = u_D^s \quad \forall Z \in \Gamma_D^s \quad \text{and} \quad \sigma^s(\varphi^s) \cdot n^s = \sigma_N^s \quad \forall Z \in \Gamma_N^s. \tag{10}$$

In addition to the above boundary conditions about the non-interface boundaries, for an FSI phenomena the fluid and structure governing equations should also satisfy the kinematic and dynamic continuity conditions along the fluid-structure interface Γ and can be written as

$$u^f(\varphi^s(Z, t), t) = u^s(Z, t) \quad \forall Z \in \Gamma, \tag{11}$$

$$\int_{\varphi^s(\gamma, t)} \sigma^f(x, t) \cdot n^f d\Gamma + \int_{\gamma} \sigma^s(Z, t) \cdot n^s d\Gamma = 0 \quad \forall \gamma \subset \Gamma, \tag{12}$$

where γ is any element on the fluid-structure interface Γ at $t = 0$ and $\varphi^s(Z, t)$ is the deformed position the material point Z at time t. In the above equation, n^f and n^s are the unit outward normal vectors to the fluid and structural domain boundaries.

As described earlier, in an ALE approach, the fluid nodes inside the fluid domain needs to be shifted so that the fluid and structural domain interior nodes do not overlap. The dynamics of fluid mesh motion can be modeled by considering a pseudo-elastic constitutive equation given by

$$\nabla \cdot \sigma^m = 0 \quad \text{where} \quad \sigma^m = (1 + \tau_m) \left[\left(\nabla \eta^f(\chi, t) + \left(\nabla \eta^f(\chi, t) \right)^T \right) + (\nabla \cdot \eta^f(\chi, t)) I \right], \tag{13}$$

satisfying the boundary conditions

$$\eta^f(Z, t) = \varphi^s(Z, t) - Z \qquad \forall \ Z \in \Gamma, \tag{14}$$

$$\eta^f(\chi, t) = 0 \qquad \forall \ \chi \in (\partial\Omega^f(0))\backslash\Gamma. \tag{15}$$

Here η^f is the displacement of the fluid mesh node and τ_m is the element-level stiffness to limit the distortion of the small elements.

3 Weak Variational Form: Combined Fluid-Structure Formulation

To construct the weak form for the fluid-structure system by introducing trial function spaces $\mathcal{S}_u$ and $\mathcal{S}_p$ and corresponding test function spaces $\mathcal{V}_u$ and $\mathcal{V}_p$ for the fluid-structure velocity and fluid pressure, respectively. The definition of the trial-and-test function spaces is as follows:

$$\mathcal{S}_u = \big\{ (\boldsymbol{u}^{\mathrm{f}}, \boldsymbol{u}^{\mathrm{s}}) | (\boldsymbol{u}^{\mathrm{f}}, \boldsymbol{u}^{\mathrm{s}}) \in H^1(\Omega^{\mathrm{f}}(t)) \times H^1(\Omega^{\mathrm{s}}),$$
$$\boldsymbol{u}^{\mathrm{f}}(\varphi^{\mathrm{s}}(\boldsymbol{Z}, t)) = \boldsymbol{u}^{\mathrm{s}}(\boldsymbol{Z}, t) \quad \forall \boldsymbol{Z} \in \Gamma,$$
$$\boldsymbol{u}(\boldsymbol{x}, t)^{\mathrm{f}} = \boldsymbol{u}_{\mathrm{D}}^{\mathrm{f}} \ \forall \boldsymbol{x} \in \Gamma_{\mathrm{D}}^{\mathrm{f}} \text{ and } \boldsymbol{u}^{\mathrm{s}}(\boldsymbol{Z}, t) = \boldsymbol{u}_{\mathrm{D}}^{\mathrm{s}} \ \forall \boldsymbol{Z} \in \Gamma_{\mathrm{D}}^{\mathrm{s}} \big\},$$
$$\mathcal{S}_p = \{ p | \, p \in L^2(\Omega^{\mathrm{f}}(t)) \}.$$

$$\mathcal{V}_u = \big\{ (\phi^{\mathrm{f}}, \phi^{\mathrm{s}}) | (\phi^{\mathrm{f}}, \phi^{\mathrm{s}}) \in H^1(\Omega^{\mathrm{f}}(t)) \times H^1(\Omega^{\mathrm{s}}),$$
$$\phi^{\mathrm{f}}(\varphi^{\mathrm{s}}(\boldsymbol{Z}, t)) = \phi^{\mathrm{s}}(\boldsymbol{Z}) \quad \forall \boldsymbol{Z} \in \Gamma,$$
$$\phi^{\mathrm{f}}(\boldsymbol{x}) = 0 \ \forall \ \boldsymbol{x} \in \Gamma_{\mathrm{D}}^{\mathrm{f}} \text{ and } \phi^{\mathrm{s}}(\boldsymbol{Z}) = 0 \ \forall \ \boldsymbol{Z} \in \Gamma_{\mathrm{D}}^{\mathrm{s}} \big\},$$
$$\mathcal{V}_p = \{ q | \, q \in L^2(\Omega^{\mathrm{f}}(t)) \}.$$

The weak form of the Navier–Stokes Eqs. (1) and (2) can be expressed as

$$\int_{\Omega^{\mathrm{f}(t)}} \rho^{\mathrm{f}} \left(\left. \frac{\partial \boldsymbol{u}^{\mathrm{f}}}{\partial t} \right|_\chi + (\boldsymbol{u}^{\mathrm{f}} - \boldsymbol{w}) \cdot \nabla \boldsymbol{u}^{\mathrm{f}} \right) \cdot \phi^{\mathrm{f}} d\boldsymbol{x} + \int_{\Omega^{\mathrm{f}(t)}} \sigma^{\mathrm{f}} : \nabla \phi^{\mathrm{f}} d\boldsymbol{x} =$$
$$\int_{\Omega^{\mathrm{f}(t)}} \boldsymbol{f}^{\mathrm{f}} \cdot \phi^{\mathrm{f}} d\boldsymbol{x} + \int_{\Gamma_{\mathrm{H}}^{\mathrm{f}}} \sigma_{\mathrm{H}}^{\mathrm{f}} \cdot \phi^{\mathrm{f}} d\Gamma + \int_{\Gamma(t)} \left(\sigma^{\mathrm{f}}(\boldsymbol{x}, t) \cdot \mathbf{n}^{\mathrm{f}} \right) \cdot \phi^{\mathrm{f}} d\Gamma, \quad (16)$$

$$\int_{\Omega^{\mathrm{f}}(t)} \nabla \cdot \boldsymbol{u}^{\mathrm{f}} q d\boldsymbol{x} = 0. \quad (17)$$

Similarly, weak form of the structural dynamics Eq. (7) can be written as

$$\int_{\Omega^{\mathrm{s}}} \rho^{\mathrm{s}} \frac{\partial \boldsymbol{u}^{\mathrm{s}}}{\partial t} \cdot \phi^{\mathrm{s}} d\boldsymbol{Z} + \int_{\Omega^{\mathrm{s}}} \sigma^{\mathrm{s}} : \nabla \phi^{\mathrm{s}} d\boldsymbol{Z} =$$
$$\int_{\Omega^{\mathrm{s}}} \boldsymbol{f}^{\mathrm{s}} \cdot \phi^{\mathrm{s}} d\boldsymbol{Z} + \int_{\Gamma_{\mathrm{H}}^{\mathrm{s}}} \sigma^{\mathrm{s}} \cdot \mathrm{H} \cdot \phi^{\mathrm{s}} d\Gamma + \int_{\Gamma} (\sigma^{\mathrm{s}}(\boldsymbol{Z}, t) \mathbf{n}^{\mathrm{s}}) \cdot \phi^{\mathrm{s}}(\boldsymbol{Z}) d\Gamma. \quad (18)$$

The weak form of the dynamic traction continuity condition in Eq. (12) will be

$$\int_{\Gamma(t)} \left(\sigma^{\mathrm{f}}(\boldsymbol{x}, t) \cdot \mathbf{n}^{\mathrm{f}} \right) \cdot \phi^{\mathrm{f}}(\boldsymbol{x}) d\Gamma + \int_{\Gamma} (\sigma^{\mathrm{s}}(\boldsymbol{Z}, t) \cdot \mathbf{n}^{\mathrm{s}}) \cdot \phi^{\mathrm{s}}(\boldsymbol{Z}) d\Gamma = 0, \quad (19)$$

where ϕ^f and ϕ^s are required to satisfy $\phi^f(\varphi^s(\cdot)) = \phi^s(\cdot)$ on Γ. A detailed derivation of the above weak form in Eq. (19) from its strong form in Eq. (12) can be found in [17]. Now we can combine Eqs. (16)–(18) using Eqs. (19) to construct a single unique relation for the combined fluid-structure domain, which is given as find $(u^f, u^s, p) \in S_u \times S_p$ such that for all $(\phi^f, \phi^s, q) \in V_u \times V_p$

$$\int_{\Omega^f(t)} \rho^f \left(\left. \frac{\partial u^f}{\partial t} \right|_{\chi} + (u^f - w) \cdot \nabla u^f \right) \cdot \phi^f(x) dx + \int_{\Omega^f(t)} \sigma^f : \nabla \phi^f dx$$

$$- \int_{\Omega^f(t)} \nabla \cdot u^f q dx$$

$$+ \int_{\Omega^s} \rho^s \frac{\partial u^s}{\partial t} \cdot \phi^s dZ + \int_{\Omega^s} \sigma^s : \nabla \phi^s dZ =$$

$$\int_{\Omega^f(t)} f^f \cdot \phi^f dx + \int_{\Gamma_H^f} \sigma_H^f \cdot \phi^f d\Gamma + \int_{\Omega^s} f^s \cdot \phi^s dZ + \int_{\Gamma_H^s} \sigma_H^s \cdot \phi^s d\Gamma. \quad (20)$$

The idea here is to solve the discrete fluid and structural domains as a single unique domain $\Omega = \Omega^f \cup \Omega^s$. In the above form, the velocity and displacement continuity conditions are satisfied implicitly. While the traction continuity condition is absorbed into the weak formulation.

4 Quasi-Monolithic Formulation

In this section, we will present a second-order time discretization of the combined fluid-structure formulation given in Sect. 3. The explicit construction of the interface at the start of each time step decouples the solid position and fluid mesh motion from the computation of fluid-structure variables (u^f, p, u^s). Additionally, the decoupling of the fluid mesh motion enables us to determine the convective velocity of the fluid flow explicitly and linearize the nonlinear Navier–Stokes relation. Hence, the quasi-monolithic formulation does not require nonlinear iteration per time step.

4.1 Second Order in Time Discretization

Let $\mathbb{P}_2(\Omega_h)$ denote the standard second-order Lagrange finite element space on domain $\Omega_h = \Omega_h^f \cup \Omega_h^s$. First, we employ the second-order extrapolation to describe the displacement vector $\eta_h^{s,n}$ of the flexible structural as

$$\eta_h^{s,n}(Z_i) = \eta_h^{s,n-1}(Z_i) + \frac{3\Delta t}{2} u_h^{s,n-1}(Z_i) - \frac{\Delta t}{2} u_h^{s,n-2}(Z_i) \quad \forall Z_i \in T_h^s. \quad (21)$$

Now that we have both the boundary conditions Eqs. (14) and (15) required for solving the ALE Eq. (13), we solve Eq. (13) employing $\mathbb{P}_1$ finite element space. The edges of an isoparametric element are assumed straight unless they are on the interface or on a curved boundaries. This assumption enables us to use $\mathbb{P}_1$ finite element instead of the $\mathbb{P}_2$ for updating the mesh. As a result of this, size of the system of linear equations required for solving the fluid mesh displacement, $\eta_h^{f,n}$, on finite element space with $\mathbb{P}_1$ discretization would be smaller than the system of linear equations required for solving the $\mathbb{P}_2$ discretization space without losing the accuracy of the coupled fluid-structure solver.

We now use the solution of $\eta_h^{f,n}$ computed on the $\mathbb{P}_1$ finite element space to update the location of triangular mesh $\mathcal{T}_{h,t^n}^f$ vertices. Since the interior edges are straight, we can position the non-vertex computational node at the center of the edge. In this way, we are able to determine the fluid mesh displacement for all the $\mathbb{P}_2$ finite element mesh $\mathcal{T}_{h,t^n}^f$ computational nodes even by solving the ALE Eq. (13) on a $\mathbb{P}_1$ finite element mesh.

The nonlinear convective term can be linearized by defining a second-order time-accurate extrapolation function given below:

$$\breve{u}_h^f(\mathbf{\Psi}_h^n(x,t^n)) = 2u_h^{f,n-1}(\mathbf{\Psi}_h^n(x,t^{n-1})) - u_h^{f,n-2}(\mathbf{\Psi}_h^n(x,t^{n-2})), \qquad (22)$$

where $\mathbf{\Psi}_h^n(\cdot,t^{n-j})$ is the backward mapping function for the spatial grid points on the mesh $\mathcal{T}_{h,t^n}^f$ to $\mathcal{T}_{h,t^{n-j}}^f$ and mesh velocity $w_h^n(x)$ is defined as

$$
\begin{aligned}
w_h^n(x) &= \sum_{i=1}^{G} \phi_i^{f,n}(x) \frac{1}{\Delta t}\left(\left(x_i^n - x_i^{n-1}\right) + \frac{1}{2}\left(x_i^{n-1} - x_i^{n-2}\right) - \frac{1}{2}\left(x_i^{n-2} - x_i^{n-3}\right)\right) \\
&= \sum_{i=1}^{G} \phi_i^{f,n}(x) \frac{1}{\Delta t}\left(x_i^n - \frac{1}{2}x_i^{n-1} - x_i^{n-2} + \frac{1}{2}x_i^{n-3}\right).
\end{aligned}
\qquad (23)
$$

Here w_h^n is a second-order approximation of the fluid mesh velocity $\partial_t \mathbf{\Psi}_h^n(x,t^n)$

$$f'(t^n) = \frac{1}{\Delta t}\left(f(t^n) - \frac{1}{2}f(t^{n-1}) - f(t^{n-2}) + \frac{1}{2}f(t^{n-3})\right) + O(\Delta t^2).$$

We next show that

$$w_h^n(x_j^n) = \breve{u}_h^f(x_j^n) \qquad \text{on } \Gamma_{h,t^n}. \qquad (24)$$

To prove the above equation, we rewrite Eq. (23) as

$$
\begin{aligned}
w_h^n(x_j^n) &= \frac{1}{\Delta t}\left((x_j^n - x_j^{n-1}) + \frac{1}{2}\left(x_j^{n-1} - x_j^{n-2}\right) - \frac{1}{2}\left(x_j^{n-2} - x_j^{n-3}\right)\right) \\
&= \frac{\varphi_h^{s,n}(Z_j) - \varphi_h^{s,n-1}(Z_j)}{\Delta t} + \frac{\varphi_h^{s,n-1}(Z_j) - \varphi_h^{s,n-2}(Z_j)}{2\Delta t} - \frac{\varphi_h^{s,n-2}(Z_j) - \varphi_h^{s,n-3}(Z_j)}{2\Delta t} \ \forall \ Z_j \in \Gamma_{h,t^n}.
\end{aligned}
\qquad (25)
$$

By substituting the definition of $\varphi^{s,n}(\boldsymbol{Z})$ from Eq. (21) into Eq. (25) and simplifying will yield

$$
\begin{aligned}
\boldsymbol{w}_{\mathrm{h}}^{n}(\boldsymbol{x}_{j}^{n}) = {} & 2\left(\frac{3}{4}\boldsymbol{u}_{\mathrm{h}}^{s,n-1}(\boldsymbol{Z}_{j}) + \frac{1}{2}\boldsymbol{u}_{\mathrm{h}}^{s,n-2}(\boldsymbol{Z}_{j}) - \frac{1}{4}\boldsymbol{u}_{\mathrm{h}}^{s,n-3}(\boldsymbol{Z}_{j})\right) \\
& - \left(\frac{3}{4}\boldsymbol{u}_{\mathrm{h}}^{s,n-2}(\boldsymbol{Z}_{j}) + \frac{1}{2}\boldsymbol{u}_{\mathrm{h}}^{s,n-3}(\boldsymbol{Z}_{j}) - \frac{1}{4}\boldsymbol{u}_{\mathrm{h}}^{s,n-4}(\boldsymbol{Z}_{j})\right).
\end{aligned}
$$

4.2 Complete Scheme

In this subsection, we present the fully discretized finite element form of the quasi-monolithic formulation. The variational statement reads as

$$
\text{find}\quad (\boldsymbol{u}_{\mathrm{h}}^{f,n}, p_{\mathrm{h}}^{f,n}, \boldsymbol{u}_{\mathrm{h}}^{s,n}) \in V_{\mathrm{h}}\left(t^{n}, \varphi_{\mathrm{h}}^{s,n}, \frac{3}{4}, \frac{1}{2}\boldsymbol{u}_{\mathrm{h}}^{s,n-1} - \frac{1}{4}\boldsymbol{u}_{\mathrm{h}}^{s,n-2}\right)
$$

with $\boldsymbol{u}_{\mathrm{h}}^{f,n}|_{\Gamma_{D}^{f}} = \boldsymbol{u}_{D}^{f}$ and $\boldsymbol{u}_{\mathrm{h}}^{s,n}|_{\Gamma_{D}^{s}} = \boldsymbol{u}_{D}^{s}$ so that for any finite element triple

$$
(\boldsymbol{\phi}^{f}, q^{f}, \boldsymbol{\phi}^{s}) \in V_{\mathrm{h}}(t^{n}, \varphi_{\mathrm{h}}^{s,n}, 1, 0) \tag{26}
$$

with $\boldsymbol{\phi}_{\mathrm{h}}^{f}|_{\Gamma_{D}^{f}} = 0$ and $\boldsymbol{\phi}_{\mathrm{h}}^{s}|_{\Gamma_{D}^{s}} = 0$, such that

$$
\begin{aligned}
\int_{\Omega_{\mathrm{h},t^{n}}^{f}} &\left[\frac{\rho^{f}}{\Delta t}\left(\frac{3}{2}\boldsymbol{u}_{\mathrm{h}}^{f,n}(\boldsymbol{x}) - 2\boldsymbol{u}_{\mathrm{h}}^{f,n-1}(\boldsymbol{\Psi}_{\mathrm{h}}^{n}(\boldsymbol{x}, t^{n-1})) + \frac{1}{2}\boldsymbol{u}_{\mathrm{h}}^{f,n-2}(\boldsymbol{\Psi}_{\mathrm{h}}^{n}(\boldsymbol{x}, t^{n-2}))\right) \right. \\
&\left. + \left(\check{\boldsymbol{u}}_{\mathrm{h}}^{f} - \boldsymbol{w}_{\mathrm{h}}^{n}\right)\cdot\nabla\boldsymbol{u}_{\mathrm{h}}^{f,n} + \frac{1}{2}\left(\nabla\check{\boldsymbol{u}}_{\mathrm{h}}^{f}\right)\boldsymbol{u}_{\mathrm{h}}^{f,n}\right]\cdot\boldsymbol{\phi}^{f}d\boldsymbol{x} \; \Big\}\; A \\
&+ \int_{\Omega_{\mathrm{h},t^{n}}^{f}} \rho^{f}\nu^{f}\left(\nabla\boldsymbol{u}_{\mathrm{h}}^{f,n} + (\nabla\boldsymbol{u}_{\mathrm{h}}^{f,n})^{T}\right):\nabla\boldsymbol{\phi}^{f}d\boldsymbol{x} \\
&\qquad\qquad - \int_{\Omega_{\mathrm{h},t^{n}}^{f}} p_{\mathrm{h}}^{f,n}(\nabla\cdot\boldsymbol{\phi}^{f})d\boldsymbol{x} \; \Big\}\; B \\
&\qquad\qquad - \int_{\Omega_{\mathrm{h},t^{n}}^{f}} q^{f}(\nabla\cdot\boldsymbol{u}_{\mathrm{h}}^{f,n})d\boldsymbol{x} \; \Big\}\; C \\
&+ \int_{\Omega_{\mathrm{h}}^{s}} \frac{\rho^{s}}{\Delta t}\left(\frac{3}{2}\boldsymbol{u}_{\mathrm{h}}^{s,n} - 2\boldsymbol{u}_{\mathrm{h}}^{s,n-1} + \frac{1}{2}\boldsymbol{u}_{\mathrm{h}}^{s,n-2}\right)\cdot\boldsymbol{\phi}^{s}d\boldsymbol{Z} \\
&+ \frac{1}{2}\int_{\Omega_{\mathrm{h}}^{s}} \left(\sigma^{s}(\varphi_{\mathrm{h}}^{s,n-1}) + \sigma^{s}(\varphi_{\mathrm{h}}^{s,n+1})\right):\nabla\boldsymbol{\phi}^{s}d\boldsymbol{Z} \; \Bigg\}\; D \\
= \int_{\Omega_{\mathrm{h},t^{n}}^{f}} &\boldsymbol{f}^{f}\cdot\boldsymbol{\phi}^{f}d\Omega + \int_{(\Gamma_{H}^{f})_{\mathrm{h}}} \boldsymbol{\sigma}_{H}^{f}\cdot\boldsymbol{\phi}^{f}d\Gamma + \int_{\Omega_{\mathrm{h}}^{s}} \boldsymbol{f}^{s}\cdot\boldsymbol{\phi}^{s}d\Omega + \int_{(\Gamma_{H}^{s})_{\mathrm{h}}} \boldsymbol{\sigma}_{H}^{s}\cdot\boldsymbol{\phi}^{s}d\Gamma, \; \Big\}\; E, \quad (27)
\end{aligned}
$$

where A and B contain the fluid velocity and pressure terms from the Navier–Stokes momentum equation, C is the Navier–Stokes continuity terms, D denotes Navier's equation for structural dynamics, and E is the right-hand side part of the combined fluid-structure system consisting of the boundary condition and external body force terms. The $\frac{1}{2}\left(\nabla \breve{u}_{\mathrm{h}}^{\mathrm{f}}\right) u_{\mathrm{h}}^{\mathrm{f,n}}$ term in part A of Eq. (27) to stabilize the convective term is introduced and discussed in detail in [22].

4.3 Algorithm

The foregoing variational formulation can be expressed in the form of an algorithm. To begin with, the details of the initial setup are as follows: there is a mesh $\mathcal{T}_{\mathrm{h}}^{\mathrm{s}}$ for the solid reference domain Ω^{s} which shares a part of its boundary grid points with $\mathcal{T}_{\Gamma}^{\mathrm{h}}$ along Γ. Assuming $u_{\mathrm{h}}^{\mathrm{f,n-1}}$, $u_{\mathrm{h}}^{\mathrm{s,n-1}}$ and $\eta_{\mathrm{h}}^{\mathrm{s}}$ are known for the mesh $\mathcal{T}_{\mathrm{h},t^{n}-1}^{\mathrm{f}}$ which is defined on the domain $\Omega_{\mathrm{h},t^{n}-1}^{\mathrm{f}}$. Here $\Omega_{\mathrm{h},t^{n}-1}^{\mathrm{f}}$ denotes the numerical approximation for the fluid domain at time t^{n-1}. It should be noted that we have considered $\mathbb{P}_{m}/\mathbb{P}_{m-1}/\mathbb{P}_{m}$ elements. To ensure the optimal rate of approximation on an isoparametric finite element mesh, all the constituent elements are considered as straight-edged standard Lagrangian elements. The basic steps to be performed in the quasi-monolithic combined fluid-structure formulation are summarized below:

Algorithm 1: Second-order quasi-monolithic formulation for fluid-structure interactions

---

1. Start with known $u_{\mathrm{h}}^{\mathrm{f,n-1}}$, $u_{\mathrm{h}}^{\mathrm{s,n-1}}$, $\eta_{\mathrm{h}}^{\mathrm{s}}$ at time t^{n-1} and t^{n-2}
2. Advance from t^{n-1} to t^{n}
   - (a) Determine the structural displacements $\eta_{\mathrm{h}}^{\mathrm{s}}$ using Eq. (21)
   - (b) Solve Eq. (13) using P_{1} elements on $\Omega_{\mathrm{h},t^{0}}^{\mathrm{f}}$ to determine mesh displacements
   - (c) Update the fluid mesh $\Omega_{\mathrm{h},t^{n}}^{\mathrm{f}}$ using the mesh displacements from (b)
   - (d) Evaluate $\breve{u}_{\mathrm{h}}^{\mathrm{f}}$ and $w_{\mathrm{h}}^{\mathrm{n}}$ by Eqs. (22) and (23)
   - (e) Solve for the updated field properties $u_{\mathrm{h}}^{\mathrm{f,n}}$, $p_{\mathrm{h}}^{\mathrm{n}}$, and $u_{\mathrm{h}}^{\mathrm{s,n}}$ at current time t^{n} using Eq. (27)

---

5 Fully Stabilized Quasi-Monolithic Formulation

One of the primary limitations of the Galerkin finite element discretization used for discretizing Ω^{f} and Ω^{s} of the quasi-monolithic formulation presented in Sect. 4 is that it will experience nonphysical spurious oscillations for convection-dominant problems [23]. Traditionally, these spurious oscillations are circumvented by replacing the traditional Galerkin method with Petrov-Galerkin methods which utilize weighting functions that have more weightage for the upstream part of the flow than the downstream [24, 25]. Such streamwise upwind techniques can be interpreted as a combination of traditional Galerkin method and a stabilization term calculated at the interior of an element. This elemental-level stabilization term introduces artificial numerical diffusion which stabilizes the spurious oscillations. The weak form of the combined fluid-structure formulation given in Eq. (20) can be written in the Galerkin/Least square (GLS) stabilization form as

$$
\left. \int_{\Omega_h^f(t)} \rho^f \left(\partial_t u^f + \left(\hat{u}^f - w \right) \cdot \nabla u^f \right) \cdot \phi^f d\Omega + \int_{\Omega^f(t)} \sigma^f : \nabla \phi^f d\Omega \right\} A
$$
$$
\left. - \int_{\Omega^f(t)} \nabla \cdot u^f q d\Omega \right\}
$$
$$
\left. + \sum_{e=1}^{n_{el}} \int_{\Omega^e} \tau_m \left[\rho^f \left(u^f - w \right) \cdot \nabla \phi^f + \nabla q \right] \cdot \right\} B
$$
$$
\left[\rho^f \partial_t u^f + \rho^f \left(u^f - w \right) \cdot \nabla u^f - \nabla \cdot \sigma^f - f^f \right] d\Omega^e \right\}
$$
$$
\left. + \sum_{e=1}^{n_{el}} \int_{\Omega^e} \nabla \cdot \phi^f \tau_c \nabla \cdot u^f d\Omega^e \right\} C
$$
$$
\left. + \int_{\Omega^s} \rho^s \partial_t u^s \cdot \phi^s d\Omega + \int_{\Omega^s} \sigma^s : \nabla \phi^s d\Omega = \right\} D
$$
$$
\left. \int_{\Omega_{h,t^n}^f} f^f \cdot \phi^f d\Omega + \int_{(\Gamma_H^f)_h} \sigma_H^f \cdot \phi^f d\Gamma + \int_{\Omega_h^s} f^s \cdot \phi^s d\Omega + \int_{(\Gamma_H^s)_h} \sigma_H^s \cdot \phi^f d\Gamma. \right\} E
$$

$$(28)$$

One can observe that terms A, D, and E combine to form the Galerkin weak form presented in Eq. (20). On the other hand, the term B represents the GLS terms for the convective and pressure to suppress the spurious oscillations for the convection-dominant problems and to circumvent the inf-sup/LBB condition, respectively. Term C denotes the stabilization term for the incompressibility constraint to provide additional stability. Unlike $\mathbb{P}_2/\mathbb{P}_1/\mathbb{P}_2$ finite element discretization for the fluid velocity, pressure and structural velocity to justify the well-posedness. The above stabilized combined fluid-structure weak form in Eq. (28) is discretized using equal order elements for both fluid velocity and pressure to simplify the computational framework significantly.

The stabilization parameters τ_m and τ_c in the term B represent the variational stabilization factors for the momentum and continuity equations [24, 26–28]. The stabilization parameter τ_m for the momentum equation is defined as [29]

$$
\tau_m = \left[\left(\frac{2\rho^f}{\Delta t} \right)^2 + (\rho^f)^2 \left(u^f - w \right) \cdot G \cdot \left(u^f - w \right) + 12(\mu^f)^2 G : G \right]^{-\frac{1}{2}}, \quad (29)
$$

where G is the elemental contravariant metric tensor which is defined as

$$G = \left(\frac{\partial \boldsymbol{\xi}}{\partial \boldsymbol{x}}\right)^T \frac{\partial \boldsymbol{\xi}}{\partial \boldsymbol{x}}, \tag{30}$$

where $\boldsymbol{\xi}$ is local element-level coordinate system and it depends on the element shape. τ_m in Eq. 29 consists of three parts, the first term represents the stabilization for the temporal dominant, second for advection dominant, and the last for diffusion-dominated cases. The stabilization factors are generally developed using the variational multiscale approach, where the finite element space is decomposed into coarse resolvable scales and fine non-resolvable scales. Therefore, the equation for non-resolvable scales forms the equation for error and the solution of this equation is approximated as the average of appropriate small-scale Green's function. This solution of the fine scale is used for determining the stabilization factors τ_m and τ_c. For more detailed mathematical treatment refer to [30] and [31]. The stabilization parameter τ_c for the continuity equation is defined as

$$\tau_c = \frac{1}{8 \operatorname{tr}(G) \tau_m}. \tag{31}$$

The fully discretized quasi-monolithic fluid-structure formulation for multiple structures using BDF2 can be written as

$$\int_{\Omega_{h,t^n}^f} \left[\frac{\rho^f}{\Delta t} \left(\frac{3}{2} u_h^{f,n}(x) - 2u_h^{f,n-1}(\Psi_h^n(x,t^{n-1})) + \frac{1}{2} u_h^{f,n-2}(\Psi_h^n(x,t^{n-2})) \right) \right.$$

$$\left. + \left(\breve{u}_h^f - w_h^n \right) \cdot \nabla u_h^{f,n} + \frac{1}{2} \left(\nabla \breve{u}_h^f \right) u_h^{f,n} \right] \cdot \phi^f dx$$

$$+ \int_{\Omega_{h,t^n}^f} \rho^f \nu^f \left(\nabla u_h^{f,n} + (\nabla u_h^{f,n})^T \right) : \nabla \phi^f dx - \int_{\Omega_{h,t^n}^f} p_h^{f,n} (\nabla \cdot \phi^f) dx$$

$$- \int_{\Omega_{h,t^n}^f} q^f (\nabla \cdot u_h^{f,n}) dx$$

$$+ \sum_{e=1}^{n_{el}} \int_{\Omega_h^e} \tau_m \left[\rho^f \left(\breve{u}_h^f - w_h^n \right) \cdot \nabla \phi^f + \nabla q \right] \cdot$$

$$\left[\frac{\rho^f}{\Delta t} \left(1.5 u_h^{f,n} - 2u_h^{f,n-1} + 0.5 u_h^{f,n-2} \right) + \rho^f \left(\breve{u}_h^f - w_h^n \right) \cdot \nabla u_h^{f,n} - \nabla \cdot \sigma^{f,n_h} - f^f \right] d\Omega^e$$

$$+ \sum_{e=1}^{n_{el}} \int_{\Omega_h^e} \nabla \cdot \phi^f \tau_c \nabla \cdot u_h^{f,n} d\Omega^e$$

$$+ \int_{\Omega_h^s} \frac{\rho^s}{\Delta t} \left(\frac{3}{2} u_h^{s,n} - 2u_h^{s,n-1} + \frac{1}{2} u_h^{s,n-2} \right) \cdot \phi^s dZ$$

$$+ \frac{1}{2} \int_{\Omega_h^s} \left(\sigma^s(\varphi_h^{s,n-1}) + \sigma^s(\varphi_h^{s,n+1}) \right) : \nabla \phi^s dZ$$

$$= \int_{\Omega_{h,t^n}^f} f^f \cdot \phi^f d\Omega + \int_{(\Gamma_H^f)_h} \sigma_H^f \cdot \phi^f d\Gamma + \int_{\Omega_h^s} f^s \cdot \phi^s d\Omega + \int_{(\Gamma_H^s)_h} \sigma_H^s \cdot \phi^f d\Gamma.$$

(32)

The implementation of the above fully stabilized quasi-monolithic combined fluid-structure formulation differs slightly from the implementation in Sect. 4. Instead of Eq. (22) we define an alternative second-order time-accurate explicit function given by

$$\check{u}_h^f(\Psi_h^n(x, t^n)) = 2.25 u_h^{f,n-1}(\Psi_h^n(x, t^{n-1})) - 1.5 u_h^{f,n-2}(\Psi_h^n(x, t^{n-2}))$$
$$+ 0.25 u_h^{f,n-3}(\Psi_h^n(x, t^{n-3})).$$

(33)

Similarly, we also define an alternate function for w_h^n as

$$w_h^n(x) = \sum_{i=1}^G \phi_i^{f,n}(x) \frac{1}{\Delta t} \left(\frac{3}{2} x_i^n - 2 x_i^{n-1} + \frac{1}{2} x_i^{n-2} \right).$$

(34)

The main reason behind redefining $\check{u}_h^f$ and w_h is that Eqs. (33)–(34) enable us to implement the exact interface continuity, i.e.,

$$u_h^{f,n} = u_h^{s,n}$$

(35)

instead of the second-order approximation in time given by Eq. (22). Unlike the velocity continuity in Eq. (11) which requires us to enforce the condition explicitly, we can satisfy the velocity continuity in Eq. (35) implicitly by treating the fluid and its corresponding solid node on the interface as a single unique node. Thereby, we can decrease the size of the algebraic system of equations required per time step compared to the implementation presented in Sect. 4.

5.1 Algorithm

Unlike the quasi-monolithic combined fluid-structure formulation in Sect. 4 where we have considered $\mathbb{P}_m/\mathbb{P}_{m-1}/\mathbb{P}_m$ elements to satisfy the inf-sup or LBB condition, here we use equal order elements for both fluid pressure and velocity. The basic steps to be performed in the fully stabilized quasi-monolithic combined fluid-structure formulation are summarized below:

Similar to the quasi-monolithic formulation presented in Sect. 4, the fully stabilized quasi-monolithic formulation also solves the combined fluid-structure system only once per time step. A matrix-free version of Krylov subspace-based iterative solvers is utilized to solve the system of equations that arise from both pseudo-elastic

---

Algorithm 2: Second-order fully stabilized quasi-monolithic formulation for fluid-structure interactions

---

1. Start with known solutions $u_h^{f,n-1}$, $u_h^{s,n-1}$, $\eta_h^{s,n-1}$ at times t^{n-1} and t^{n-1}
2. Advance from t^{n-1} to t^n
   (a) Determine the structural displacements $\eta_h^{s,n}$ using Eq. (21)
   (b) Solve Eq. (13) on Ω_{h,t^0}^f to determine mesh displacements
   (c) Update the fluid mesh Ω_{h,t^n}^f using the mesh displacement from (b)
   (d) Evaluate $\breve{u}_h^f$ and w_h^n by Eqs. (33) and (34)
   (e) Determine the element level stabilization parameters τ_m and τ_c using Eqs. (29) and (31) respectively.
   (f) Solve for the updated field properties $u_h^{f,n}$, p_h^n, and $u_h^{s,n}$ at current time t^n using Eq. (32)

---

mesh motion and combined fluid-structure equations. To scale the fluid-structure solver for large-scale computations using distributed memory parallel cluster, we next present the parallel finite element implementation of three-dimensional incompressible flow interacting with generic elastic structures for high Re flow.

6 Conclusions

In this work, we have discussed about two different ALE-based improvised monolithics, i.e., quasi-monolithic, FSI formulations that are computationally efficient and numerical stable for low mass ratios. In both these formulations, the fluid mesh motion has been decoupled from the monolithic matrix consisting of governing equations that describe the fluid flow, structural dynamics, interface conditions, and the mesh motion. As a result, the size of the matrix which needs to be solved reduces by a maximum of 20% in the case of two-dimensional simulations and a maximum of 40% for the three-dimensional simulations. The decoupling of mesh motion has been made possible by predicting the structural displacements at the start of each time step based on the previous time step velocities. Additionally, both these methods linearize the convective velocities by using a second-order explicit approximation based on previous time step information. The first quasi-monolothic approach discussed uses an extra stabilization term, $\frac{1}{2}\left(\nabla \breve{u}_h^f\right) u_h^{f,n}$, which has been proposed by Temam to provide numerical stability. This term plays a role in proving the unconditional stability of the method theoretically. The formulation is stable for any mixed finite element discretization for the velocity and pressure. On the other hand, the second approach discussed in this paper considers a Galerkin least square-based stabilization to provide convective stabilization for convectively dominant problems. Additionally, this method requires equal order finite element for the fluid velocity and pressure. As a result, this approach would require lower memory requirements compared to the first approach.

References

1. Paidoussis M, Price SJ, De Langre E (2010) Fluid-structure interactions: cross-flow induced instabilities. Cambridge University Press
2. Naudascher E, Rockwell D (2005) Flow-induced vibrations: an engineering guide. Dover, New York
3. Gurugubelli PS, Jaiman R (2015) Self-induced flapping dynamics of a flexible inverted foil in a uniform flow. J Fluid Mech 781:657–694
4. Blevins RD (1990) Flow-induced vibration. Van Nostrand Reinhold, New York
5. Gurugubelli PS, Ghoshal R, Joshi V, Jaiman RK (2018) A variational projection scheme for nonmatching surface-to-line coupling between 3d flexible multibody system and incompressible turbulent flow. Comput Fluids 165:160–172
6. Joshi V, Gurugubelli PS, Law YZ, Jaiman RK, Adaikalaraj PFB (2018) A 3D coupled fluid-flexible multibody solver for offshore vessel-riser system, OMAE2018-78281. In: ASME 2018 37th international conference on ocean, offshore and arctic engineering
7. Jaiman RK, Sen S, Gurugubelli P (2015) A fully implicit combined field scheme for freely vibrating square cylinders with sharp and rounded corners. Comput Fluids 112:1–18
8. Arienti M, Hung P, Morano E, Shepherd E (2003) A level set approach to Eulerian-Lagrangian coupling. J Comput Phys 185:213–251
9. Peskin CS (2002) The immersed boundary method. Acta Numerica 11:479–517
10. Glowinski R, Pan TW, Periaux J (1994) A fictitious domain method for external incompressible viscous flow modeled by Navier-Sstokes equations. Comput Methods Appl Mech Eng 112:133–148
11. Donea J, Giuliani S, Halleux JP (1982) An arbitrary lagrangian-eulerian finite element method for transient dynamic fluid-structure interactions. Comput Methods Appl Mech Eng 33:689–723
12. Farhat C, Lesoinne M, Maman N (1995) Mixed explicit/implicit time integration of coupled aeroelastic problems: three field formulation, geometric conservation and distributed solution. Int J Numer Methods Eng 21:807–835
13. Felippa CA, Park KC, Farhat C (2001) Partitioned analysis of coupled mechanical systems. Comput Methods Appl Mech Eng 190:3247–3270
14. Causin P, Gerbeau JF, Nobile F (2005) Added-mass effect in the design of partitioned algorithms for fluid-structure problems. Comput Methods Appl Mech Eng 194:4506–4527
15. Matthies HG, Niekamp R, Steindorf J (2006) Algorithms for strong coupling procedures. Comput Methods Appl Mech Eng 195:2028–2049
16. Hron J, Turek S (2006) A monolithic fem/multigrid solver for an ale formulation of fluid-structure interaction with applications in biomechanics. In: 146–170, editor, Fluid-structure interaction. Springer
17. Liu J (2012) One field formulation and a simple explicit scheme for fluid structural interaction
18. Liu J, Jaiman RK, Gurugubelli PS (2014) A stable second-order scheme for fluid-structure interaction with strong added-mass effects. J Comput Phys 270:687–710
19. Gurugubelli PS, Jaiman RK (2019) Large amplitude flapping of an inverted elastic foil in uniform flow with spanwise periodicity. J Fluids Struct 90:139–163
20. Temam R (2005) Mathematical modeling in continuum mechanic. Cambridge Publishers
21. Antman SS (1995) Nonlinear problems of elasticity. Springer
22. Temam R (2001) Navier-Stokes equations. AMS Chelsea Publishing, Theory and numerical analysis
23. Donea J, Huerta A (2003) Finite element methods for flow problems. Wiley
24. Brooks AN, Hughes TJR (1982) Streamline upwind/petrov-galerkin formulation for convection dominated flows with particular emphasis on the incompressible navier-stokes equations. Comput Methods Appl Mech Eng 32:199–259
25. Hughes TJR (1987) The finite element method. Prentice-Hall, Englewood Cliffs, NJ

26. Shakib F, Hughes T, Johan Z (1989) A new finite element formulation for computational fluid dynamics.x. the compressible euler and navier-stokes equations. Comput. Methods Appl Mech Eng 73:173–189
27. Hughes T, Franca L, Hülbert G (1989) A new finite element formulation for computational fluid dynamics. viii. The Galerkin/east-squares method for advective-diffusive equations. Comput Methods Appl Mech Eng 73:173–189
28. Franca L, Frey SL (1992) Stabilized finite element methods: II. The incompressible navier-stokes equations. Comput Methods Appl Mech Eng 99:209–233
29. Akkerman I, Bazilevs Y, Benson D, Farthing MW, Kees CE (2012) Free-surface flow and fluid-object interaction modeling with emphasis on ship hydrodynamics. J Appl Mech 79(010905)
30. Hughes TJR (1995) Multiscale phenomena: Green's functions, the dirichlet-to-neumann formulation, subgrid scale models, bubbles and the origins of stabilized methods. Comput Methods Appl Mech Eng 127:387–401
31. Codina R (2002) Stabilized finite element approximation of transient incompressible flows using orthogonal subscales. Comput Methods Appl Mech Eng 191:4295–4321

Multi-Phase Fluid-Structure Interaction with Diffused Interface Capturing

Vaibhav Joshi and Pardha S. Gurugubelli

Abstract Two-phase fluid-structure interactions are commonly observed phenomena in industries as well as daily life ranging from small-scale droplet interactions to large-scale ocean wave-current interaction with offshore marine structures. These applications involve complex multiscale and coupled nonlinear dynamics which can be quite challenging to predict and analyze via physical experiments and theoretical ways. We form a three-dimensional numerical framework based on variational finite element methods for such complex problems. The flow is modeled using incompressible Navier–Stokes equations while the fluid-fluid two-phase interface is evolved by solving phase-field Allen-Cahn equation, both written in an arbitrary Lagrangian-Eulerian (ALE) setting. The structure is modeled by the multibody structural equation and the fluid-structure interface is tracked exactly by moving grid. Thus, the formulation gives a hybrid Allen-Cahn/ALE scheme having the feature of sharp-interface tracking for the fluid-structure interface and interface capturing for the fluid-fluid interface. The coupling is carried out in a partitioned block-iterative manner which ensures flexibility and ease of implementation in the 3D parallel variational solver. The study is concluded by demonstrating the practical application of the drilling vessel-riser system with turbulent ocean current, wind streams, and free-surface waves.

V. Joshi (✉)
Department of Mechanical Engineering, Birla Institute of Technology & Science Pilani,
K K Birla Goa Campus, Goa, India
e-mail: vaibhavj@goa.bits-pilani.ac.in

P. S. Gurugubelli
Department of Mechanical Engineering, Birla Institute of Technology & Science Pilani,
Hyderabad Campus, Hyderabad, India
e-mail: pardhasg@hyderabad.bits-pilani.ac.in

R. Srinivas et al. (eds.), *Advances in Computational Modeling and Simulation*,
Lecture Notes in Mechanical Engineering,
https://doi.org/10.1007/978-981-16-7857-8_13

1 Introduction

Two-phase fluid-structure interaction (FSI) finds its applications in offshore pipelines which carry two-phase oil-gas mixture [1, 2], marine ships/vessels which are subjected to free-surface ocean waves, hemodynamics, and flow through heat exchangers, among others. The present study deals with the offshore drilling process where an offshore vessel drills the ocean floor via long slender pipelines known as risers to extract natural resources. These pipelines are subjected to turbulent ocean currents which may lead to fluid-elastic instabilities such as vortex-induced vibrations (VIV) [3, 4], while the vessel is subjected to the free-surface ocean waves. This complex drilling vessel-riser system when exposed to adverse ocean environments may lead to failure of the structure and operational delay due to the nonlinear effects of FSI. Therefore, it is imperative to study the coupled FSI problem to mitigate such circumstances of failure and provide better engineering designs.

Fundamentally, two-phase FSI comprises nonlinear interactions at the interfaces, viz., fluid-structure interface and fluid-fluid interface. Some of the challenges are the satisfaction of the no-slip condition at the fluid-structure interface, the deformation of the fluid-fluid interface with time, the accurate tracking of the structure as it deforms, and the satisfaction of equilibrium conditions at the fluid-structure interface. The boundary conditions at the fluid-structure interface can either be modeled by a fictitious force field using immersed boundary approach [5, 6], or it can be tracked exactly with the help of conforming body-fitted moving mesh approach by arbitrary Lagrangian-Eulerian (ALE) description [7], which is advantageous in accurate modeling of the boundary layer characteristics and near-wall turbulence.

On the other hand, the fluid-fluid interface between the two fluid phases can be represented by two techniques: interface-tracking (front tracking [8], particle tracking [9], ALE) and interface-capturing (level-set [10], volume-of-fluid [11], phase field). The former technique requires remeshing and other numerically expensive corrections in case of large topological changes of the fluid-fluid interface such as merging and breaking. However, interface-capturing implicitly captures the interface on a fixed Eulerian mesh. Level-set and volume-of-fluid (VOF) methods further require geometric manipulations such as re-initialization and reconstruction, respectively, to evaluate the curvature of the interface, making them tedious to implement in three dimensions for generic unstructured meshes on complex geometries. The diffused interface phase-field method assumes a gradual and smooth variation of the physical properties across the interface of a finite thickness. The physical properties such as density and viscosity of the fluid phases gradually transit across the diffused interface as a function of phase indicator or order parameter, which is solved by the minimization of free-energy functional [12]. The phase-field method does not require any geometric manipulation at the fluid-fluid interface and the curvatures to model the capillary/surface tension effects are evaluated implicitly. Moreover, the mass conservation can be imposed in a relatively simple manner, in contrast to the level-set approach. As the mesh is fixed, topological changes in the interface can also be captured to sufficient accuracy. Therefore, phase-field method for the fluid-fluid

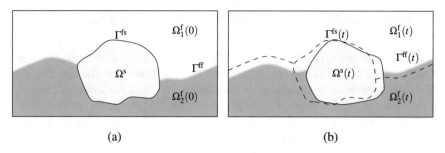

Fig. 1 Schematic of the two-phase fluid-structure interaction depicting **a** the initial undeformed state at $t = 0$, and **b** deformed configuration of the structure at time $t > 0$. Here, $\Omega^f(0)$ and $\Omega^f(t)$ denote the fluid domain at the different configurations, and Ω^s, $\Omega^s(t)$ represent the structural domain at the initial and the deformed configurations, respectively. The fluid-structure interface Γ^{fs} is assumed sharp, while the fluid-fluid interface Γ^{ff} is diffused and has a finite thickness

interface along with the exact mesh movement (ALE) of the fluid-structure interface offers advantages with regard to two-phase FSI modeling. This forms the numerical treatment of the interfaces for the present study, which has been summarized in Fig. 1.

The coupling of the different physical fields in FSI can be carried out by either monolithic [13–15] or partitioned approach [16–19]. Although robust and stable for low structure-to-fluid mass ratios, the monolithic approach does not offer flexibility and modularity in the numerical implementation, which are very essential when dealing with such complex large-scale coupled system consisting of multiple fields, namely, fluid, structure, mesh, phase field, and turbulence. In a partitioned approach, the different fields are solved in a sequential manner, while data across the different interfaces is exchanged. Partitioned methods suffer from convergence issues in the low structure-to-fluid mass ratio regimes [20–22], typically found in offshore applications. Therefore, special treatment for accelerating the convergence is required [20, 23, 24] which consists of nonlinear iterations to correct the fluid forces transferred from the fluid to the structural domain. Using a partitioned approach for such complex coupling gives some attractive advantages with regard to iterative solvers, preconditioning strategies, scalability, and parallel processing.

In the current work, we follow the partitioned strategy to model the drilling vessel-riser system where the vessel is subject to two-phase free-surface ocean waves, the riser interacts with turbulent ocean current, and the vessel-riser system is represented as a multibody structural system. We begin with a brief overview of the governing equations involved in the model, leading to the partitioned two-phase FSI solver in Sect. 2. We then perform numerical computations pertaining to internal two-phase flow in a riser in Sect. 3, followed by validation for the free-surface flow over a Wigley hull in Sect. 4 and finally demonstrate the drilling vessel-riser system in Sect. 5. We conclude the work by providing a summary and key findings in Sect. 6.

2   Two-Phase Fluid-Structure Interaction Formulation

We briefly discuss the governing equations involved in the two-phase turbulent fluid-structure interaction via a hybrid ALE/phase-field methodology. The discretization of the equations has not been discussed here for brevity. Then, the partitioned strategy for the coupling of the different equations is summarized for the formulation.

2.1   The Governing Equations

The Two-Phase Flow Equations

Consider a d-dimensional spatial fluid domain $\Omega^{\mathrm{f}}(t) \subset \mathbb{R}^d$ with a piecewise smooth boundary $\Gamma^{\mathrm{f}}(t)$. The domain $\Omega^{\mathrm{f}}(t)$ consists of two immiscible, incompressible, and Newtonian fluid phases occupying the sub-domains $\Omega_1^{\mathrm{f}}(t)$ and $\Omega_2^{\mathrm{f}}(t)$ with a fluid-fluid interface $\Gamma^{\mathrm{ff}}(t)$ between them. The governing equations for the one-fluid formulation for a viscous, incompressible, and immiscible two-phase system in the ALE referential coordinate system χ are given by the Navier–Stokes equations as

$$\rho^{\mathrm{f}} \frac{\partial \boldsymbol{v}^{\mathrm{f}}}{\partial t}\bigg|_{\chi} + \rho^{\mathrm{f}}(\boldsymbol{v}^{\mathrm{f}} - \boldsymbol{w}) \cdot \nabla \boldsymbol{v}^{\mathrm{f}} = \nabla \cdot \boldsymbol{\sigma}^{\mathrm{f}} + \nabla \cdot \boldsymbol{\sigma}^{\mathrm{turb}} + \mathbf{sf} + \rho^{\mathrm{f}} \boldsymbol{b}^{\mathrm{f}}, \quad \text{on } \Omega^{\mathrm{f}}(t) \times [0, T], \quad (1)$$

$$\nabla \cdot \boldsymbol{v}^{\mathrm{f}} = 0, \qquad\qquad\qquad \text{on } \Omega^{\mathrm{f}}(t) \times [0, T], \quad (2)$$

where $\boldsymbol{v}^{\mathrm{f}}$ and $\boldsymbol{w}$ represent the fluid and the mesh velocities, respectively, defined for each spatial point $\boldsymbol{x}^{\mathrm{f}} \in \Omega^{\mathrm{f}}(t)$, ρ^{f} is the fluid density, $\mathbf{sf}$ denotes the surface tension singular force replaced by the continuum surface force in the diffused interface description, and $\boldsymbol{b}^{\mathrm{f}} = \boldsymbol{g}$ is the body force applied on the fluid such as the gravitational force, with $\boldsymbol{g}$ being the acceleration due to gravity. The Cauchy stress tensor is denoted by $\boldsymbol{\sigma}^{\mathrm{f}} = -p\boldsymbol{I} + \mu^{\mathrm{f}}(\nabla \boldsymbol{v}^{\mathrm{f}} + (\nabla \boldsymbol{v}^{\mathrm{f}})^T)$, where p is the fluid pressure and μ^{f} is the dynamic viscosity of the fluid. The turbulent stress term is represented by $\boldsymbol{\sigma}^{\mathrm{turb}}$. The density and viscosity are dependent on the phase indicator function ϕ (known as the order parameter) as $\rho^{\mathrm{f}}(\phi) = \frac{1+\phi}{2}\rho_1^{\mathrm{f}} + \frac{1-\phi}{2}\rho_2^{\mathrm{f}}$ and $\mu^{\mathrm{f}}(\phi) = \frac{1+\phi}{2}\mu_1^{\mathrm{f}} + \frac{1-\phi}{2}\mu_2^{\mathrm{f}}$, respectively, where ρ_i^{f} and μ_i^{f} are the density and dynamic viscosity of the ith phase of the fluid, respectively. The fluid-fluid interface is evolved by solving the phase-field equation for the order parameter. The flow equations are discretized by stabilized finite element variational formulation, the details of which can be found in [24, 25].

The Allen-Cahn Equation

The order parameter ϕ which indicates the fluid-fluid interface is evolved by phase-field equation. Here, we utilize the second-order Allen-Cahn equation [26] which minimizes the Ginzburg-Landau free-energy functional $\mathcal{E}(\phi) = \int_{\Omega^{\mathrm{f}}(t)} \left(\frac{\varepsilon^2}{2}|\nabla\phi|^2 + F(\phi)\right)d\Omega$, where the first and the second terms denote the interfacial and the bulk energy of the two-phase fluid system , respectively. Here, $F(\phi) = \frac{1}{4}(\phi^2 - 1)^2$ is a double-well potential function having two minima ($\phi = 1$ and $\phi = -1$) corresponding to the two stable phases, and ε is a measure of the thickness of the diffused interface. The mass conservative Allen-Cahn equation in the moving mesh ALE framework (to account for the moving fluid-structure interface) is given by [25, 27]

$$\left.\frac{\partial\phi}{\partial t}\right|_{\chi} + (\boldsymbol{v}^{\mathrm{f}} - \boldsymbol{w}) \cdot \nabla\phi - \gamma\big(\varepsilon^2\nabla^2\phi - F'(\phi) + \beta(t)\sqrt{F(\phi)}\big) = 0, \qquad \text{on } \Omega^{\mathrm{f}}(t) \times [0, T], \quad (3)$$

where γ is a mobility parameter selected as 1 for simplicity, $F'(\phi)$ is the derivative of the potential function with respect to ϕ, and $\beta(t)$ is the time-dependent component of the Lagrange multiplier term for imparting the mass conservation property [25, 27].

As the fluid-fluid interface involves high gradients of the physical properties such as density, viscosity, and pressure across the two phases, it is a numerical challenge to sufficiently resolve these gradients and avoid any spurious unphysical oscillations in the numerical solution. In the present scenario, we have utilized the recently proposed positivity-preserving variational (PPV) scheme [25, 28, 29].

The Fluid-Fluid Interface
Equilibrium at the fluid-fluid interface of the two phases requires the satisfaction of the pressure-jump and the velocity continuity conditions. In the diffused interface description, these conditions are replaced by a continuum surface force (CSF) [30] depending on the order parameter. Thus, the surface tension force in Eq. (1) can be written in several forms reviewed in [31, 32]. Here we consider $\mathbf{sf}(\phi) = \sigma\varepsilon\alpha_{\mathrm{sf}}\nabla \cdot (|\nabla\phi|^2\boldsymbol{I} - \nabla\phi \otimes \nabla\phi)$, where $\alpha_{\mathrm{sf}} = 3\sqrt{2}/4$ is a constant and σ is the surface tension coefficient between the two fluid phases.

The Turbulence Equation
The turbulence stress term in Eq. (1) is represented by a hybrid RANS/LES model which relies on Spalart-Allmaras-based delayed detached eddy simulation. The stress term is given by $\boldsymbol{\sigma}^{\mathrm{turb}} = \mu_T^{\mathrm{f}}(\nabla\boldsymbol{v}^{\mathrm{f}} + (\nabla\boldsymbol{v}^{\mathrm{f}})^T)$, where μ_T^{f} denotes the turbulent dynamic viscosity, which is computed by solving for turbulent eddy viscosity $\tilde{\nu}$ in the ALE moving mesh framework as

$$\left.\frac{\partial\tilde{\nu}}{\partial t}\right|_{\chi} + (\boldsymbol{v}^{\mathrm{f}} - \boldsymbol{w}) \cdot \nabla\tilde{\nu} = P - D + \frac{1}{\tilde{\sigma}}[\nabla \cdot ((\nu^{\mathrm{f}} + \tilde{\nu})\nabla\tilde{\nu}) + c_{b2}(\nabla\tilde{\nu})^2], \qquad \text{on } \Omega^{\mathrm{f}}(t) \times [0, T],$$

$$(4)$$

where $D = c_{w1}f_w(\tilde{\nu}/\tilde{d})^2$ and $P = c_{b1}\tilde{S}\tilde{\nu}$ denote the destruction and production terms; $\tilde{S} = S + (\tilde{\nu}/(\kappa^2\tilde{d}^2))f_{v2}$, S being the magnitude of vorticity; $\tilde{d}$ denoting a function depending on the distance to the closest wall; and c_{b1}, c_{b2}, $\tilde{\sigma}$, κ, c_{w1}, and f_w have been defined in [33]. The turbulence equation is also discretized using the stabilized finite element formulation employing the PPV scheme [34, 35].

The Multibody Structural Equation
The two-phase fluid interacts with a multibody structural system, thus resulting in the deformation of the structure. For modeling the structure, consider a d-dimensional structural domain of a component i in a multibody system $\Omega_i^{\mathrm{s}} \subset \mathbb{R}^d$ with a piecewise

smooth boundary Γ_i^s at time $t = 0$, consisting of the material coordinates X. Subject to the fluid forces, the structure deforms to a new configuration $\Omega_i^s(t)$ at time t. Let $\varphi(X, t) : \Omega_i^s \to \Omega_i^s(t)$ be a one-to-one mapping (denoting the position vector) between the material coordinates X at $t = 0$ to its position in $\Omega_i^s(t)$. The structural equation for a multibody component i is given as

$$\rho_i^s \frac{\partial^2 \varphi}{\partial t^2} = \nabla \cdot \sigma^s + \rho_i^s b^s, \qquad \text{on } \Omega_i^s, \qquad (5)$$

where ρ_i^s and b^s denote the density and the body force on the structural component i, respectively. The quantity σ^s is the first Piola-Kirchhoff stress tensor which can be expressed as a function of Cauchy-Green-Lagrangian strain tensor $E(u^s) = (1/2)[(I + \nabla u^s)^T (I + \nabla u^s) - I]$, where u^s is the displacement field for the structure, i.e., $u^s = \varphi(X, t) - X$. Furthermore, the structural velocity is expressed as $v^s = \partial \varphi / \partial t = \partial u^s / \partial t$. Further details can be found in [34, 36, 37].

The Fluid-Structure Interface

The fluid-structure interface is treated as a sharp interface in the present description and is tracked exactly via the moving mesh ALE framework. For the coupling between the fluid and the structural fields, the kinematic and dynamic equilibrium conditions are required to be satisfied at the fluid-structure interface. Let Γ^{fs} denote the fluid-structure interface at $t = 0$, i.e., at the undeformed configuration of the structure. The conditions can be written mathematically as

$$v^f(\varphi(X, t), t) = v^s(X, t), \qquad \forall X \in \Gamma^{fs}, \quad (6)$$

$$\int_{\varphi(\widehat{\gamma}, t)} \sigma^f(x^f, t) \cdot n^f d\Gamma + \int_{\widehat{\gamma}} \sigma^s(X, t) \cdot n^s d\Gamma = 0, \qquad \forall \widehat{\gamma} \subset \Gamma^{fs}, \quad (7)$$

where n^f and n^s denote the outward unit normals to the fluid and the structural domains, respectively, $\widehat{\gamma}$ is any part of the interface Γ^{fs}, and $\varphi(\widehat{\gamma}, t)$ denotes the corresponding fluid part of the interface at time t. The continuity of velocity across the interface is ensured by Eq. (6) and the balance of tractions across the interface is depicted by Eq. (7).

The Mesh Equation

Apart from the two fluid-structure interface conditions described above, a displacement continuity condition is also satisfied at the fluid-structure interface, where the displacement of the mesh nodes is equated with that of the structural displacement at the interface. Let u^f denote the mesh displacement field of the fluid domain mesh. The displacement continuity condition is thus written as $u^f = u^s$, on Γ^{fs}. For the motion of the fluid nodes apart from the fluid-structure interface, the equation $\nabla \cdot \sigma^m = 0$ is solved, where σ^m denotes the stress experienced by the fluid mesh due to the strain as a result of the structural motion. Assuming the fluid mesh to be linearly elastic, $\sigma^m = (1 + \tau_{mesh})[\nabla u^f + (\nabla u^f)^T + (\nabla \cdot u^f)I]$, where τ_{mesh} represents a mesh stiffness parameter which can be adjusted to check the distortion of the small elements.

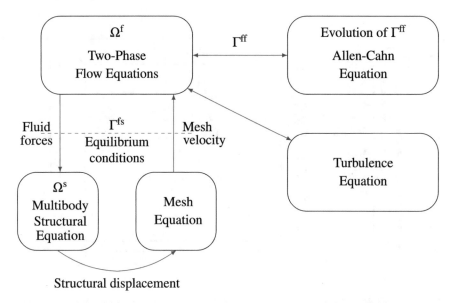

Fig. 2 Schematic of the partitioned iterative coupling of the two-phase turbulent fluid-structure interaction formulation

The fluid mesh velocity can thus be evaluated as $\boldsymbol{w} = \partial \boldsymbol{u}^f / \partial t$ and forms the closure for the moving mesh framework in the two-phase flow equations.

2.2 The Partitioned Iterative Coupling

In this section, we present the nonlinear partitioned iterative coupling between the different fields of equations involved in the two-phase turbulent fluid-structure interaction formulation. The schematic of the coupling is shown in Fig. 2.

Each field equation is solved separately and the required data is transferred between the respective fields in a sequential manner. Numerical FSI instabilities pertaining to low structure-to-fluid mass ratios are resolved by nonlinear iterative force correction (NIFC) scheme involving corrections of the fluid forces via nonlinear iterations [20, 23, 24]. This technique increases stability and convergence of the FSI solver. Furthermore, the partitioned strategy is advantageous in providing flexibility and ease in numerical implementation and coding. The linear system of equations is solved by the generalized minimal residual (GMRES) algorithm [38]. The coupled two-phase FSI system has been implemented to have the benefits of hybrid parallelism [39] for parallel computing, consisting of the state-of-the-art hierarchical memory and parallel architectures.

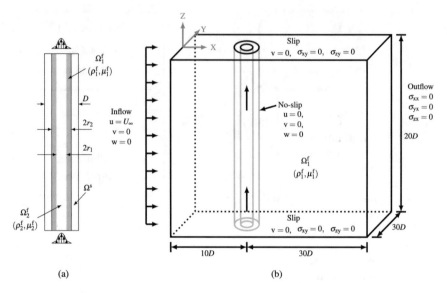

Fig. 3 Schematic of the FSI of a flexible riser with internal two-phase flow: **a** the $X - Z$ cross section of the riser with the velocity profile for the two-phase flow, and **b** the computational domain and boundary conditions

The presented partitioned coupling for the modeling of two-phase turbulent fluid-structure interaction has been validated extensively for several benchmark tests in [24, 25, 29, 34, 40], and thus we do not present those numerical tests here for brevity.

3 FSI of a Flexible Riser with Internal Two-Phase Flow

In this section, we present the application of the two-phase FSI framework to a practical problem of a riser (with internal two-phase flow) subjected to external uniform current flow, which is typically observed during the drilling process at an offshore site. The schematic of the problem statement is depicted in Fig. 3. The riser is selected as a concentric pipe with outer diameter of D and aspect ratio $L/D = 20$. A uniform freestream velocity of $v^f = (U_\infty, 0, 0)$ interacts with the outer surface of the riser, whereas a two-phase flow (consisting of the phases Ω_1^f and Ω_2^f) exists inside the riser. For the internal two-phase flow, a concentric velocity profile is prescribed at the inlet and outlet of the riser, as shown in Fig. 3a. The profile of the velocity is such that the flow entering the riser is co-annular, fully developed, and laminar [41]. The parameters considered for the problem are $r_1 = 0.2$, $r_2 = 0.4$ and $g = (0, 0, 0)$, i.e., no acceleration due to gravity. To model the riser structure, we utilize the modal analysis of the Euler-Bernoulli beam equation. The non-dimensional parameters for the VIV of the riser with the

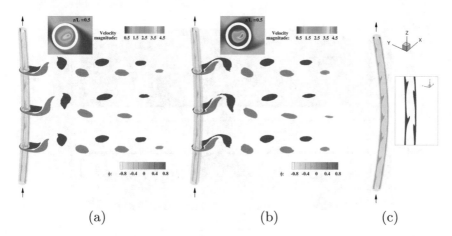

Fig. 4 FSI of flexible riser with internal two-phase flow at tU_∞/D: **a** 80, **b** 90 and **c** the internal two-phase flow pattern along the riser. The inset figure shows the magnitude of velocity at the mid-point of the riser span

internal flow are the Reynolds number $Re = (\rho_1^f U_\infty D)/\mu_1^f = (\rho_2^f U_\infty D)/\mu_2^f = 100$, mass ratio $m^* = m^s/(\pi D^2 L \rho_1^f/4) = 2.89$, density ratio $\rho^* = \rho_1^f/\rho_2^f = 100$, viscosity ratio $\mu^* = \mu_1^f/\mu_2^f = 100$, reduced velocity $U_r = U_\infty/(f_1 D) = 5$, dimensionless axial tension $P^* = P/(\rho_1^f U_\infty^2 D^2) = 0.34$, dimensionless flexural rigidity $EI^* = (EI)/(\rho_1^f U_\infty^2 D^4) = 5872.8$, and $\rho^s/\rho_1^f = 6.68$. Here, f_1 denotes the first eigenmode frequency of the riser.

The amplitude of the riser is found to be maximum at the mid-point along its span and a standing wave-like pattern is observed along the riser, both in the in-line (X) and the cross-flow (Y) directions. The Z-vorticity contours along the span of the riser are shown in Fig. 4 where the internal two-phase flow is also visualized with the help of order parameter ϕ. The topological evolution of the two-phase fluid-fluid interface is captured qualitatively, where a transition in the two-phase flow pattern is observed. Such changes in flow patterns due to VIV are crucial for improving multiphase flow assurance in these pipelines.

4 Flow Across a Wigley Hull

Before demonstrating the solver for a drillship-riser system, we validate the free-surface effects on a simplified parabolic Wigley hull. The length, beam, and draft of the hull are $L = 4$ m, $B = 0.4$ m, and $D = 0.25$ m, respectively. The computational domain with the initial free-surface position at $z = 0$ m is shown in Fig. 5. We study the wave pattern on the hull at two Froude numbers $Fr = U_\infty/\sqrt{|g|L}$, viz., 0.25 and 0.316, where U_∞ is the freestream velocity of the incoming flow and $g = (0, 0, -9.81)$ is the acceleration due to gravity. Systematic mesh convergence

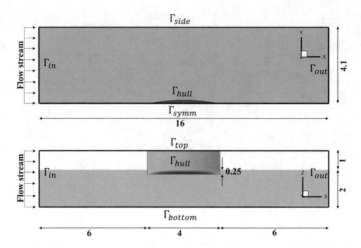

Fig. 5 Computational domain for free-surface flow across a parabolic Wigley hull

study is conducted by considering three meshes with increasing refinement. Meshes 1, 2, and 3 consist of 0.38 million, 2.4 million, and 4.6 million nodes, respectively. The wave profiles along the hull surface are measured and compared with the experimental results [42] for two Froude numbers in Fig. 6 where a good agreement is found.

The contours of the free-surface colored by the wave elevation for $Fr = 0.25$ are shown in Fig. 7 where we observe the formation of the Kelvin waves near the downstream of the hull. Pertaining to the symmetry of the hull along the $X - Z$-plane, the contours are mirrored to visualize the Kelvin waves on both sides of the hull.

5 Coupled Drillship-Riser System

We next demonstrate the developed numerical formulation for a more complicated geometry of the hull and analyze the effect of the free-surface ocean waves and turbulent current on the coupled motion of the ship-riser system. The ship hull is considered as the geometry of Navy surface combatant model DTMB 5415 [43] with a scale of 1:1. This corresponds to a length between perpendiculars of $L_{pp} = 142$ m with a draft of 6.15 m and displaced volume of water of 8424.4 m^3. A riser of diameter $D = 1$ m and aspect ratio $L/D = 200$ is attached to the bottom of the vessel.

We present the response of the vessel and the riser which are subjected to the combined effects of turbulent ocean current and the free-surface ocean waves, as shown in Fig. 8a. The computational mesh is also depicted in Fig. 8b. The mesh is refined near the fluid-fluid interface (free-surface). The free-surface waves are generated by prescribing second-order nonlinear Stokes waves at the inlet boundary. The ship is

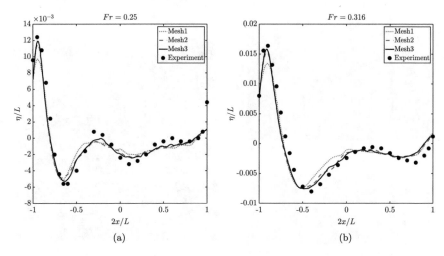

Fig. 6 Wave profile elevation along the Wigley hull: comparison between the experimental data and numerical results at **a** $Fr = 0.25$ and **b** $Fr = 0.316$. Here, η denotes the elevation of the free surface along the surface of the hull

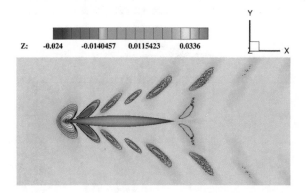

Fig. 7 Wave elevation contour along the Wigley hull at $Fr = 0.25$

modeled as a rigid body having three translational degrees of freedom, whereas the riser is solved as a nonlinear flexible beam. The connection between the ship and the riser is assumed to be rigid, i.e., the response of the ship gets completely transferred to the riser and *vice versa*. The non-dimensional parameters employed for the demonstration can be summarized as: $Re = \rho_1^f U_\infty L_{pp}/\mu_1^f = 1.42 \times 10^8$, $\rho^* = \rho_1^f/\rho_2^f = 816$, $\mu^* = \mu_1^f/\mu_2^f = 50$, $Fr = U_\infty/(\sqrt{|g|L_{pp}}) = 0.027$, $m_{ship}^* = m_{ship}^s/(\rho_1^f V_{disp}) = 1.0$, $m_{riser}^* = m_{riser}^s/(\rho_1^f D^2 L\pi/4) = 2.23$, $EI^* = EI/(\rho_1^f U_\infty^2 D^4) = 2.1158 \times 10^7$, $EA^* = EA/(\rho_1^f U_\infty^2 D^2) = 1.7706 \times 10^8$, and $U_r = U_\infty/(f_1 D) = 7.33$, where m_{ship}^* and m_{riser}^* are the mass ratios of the ship and the riser, respectively, EI^* and EA^* are the non-dimensional flexural and axial rigidity, respectively, and U_r is the reduced velocity of the riser assuming its first natural frequency.

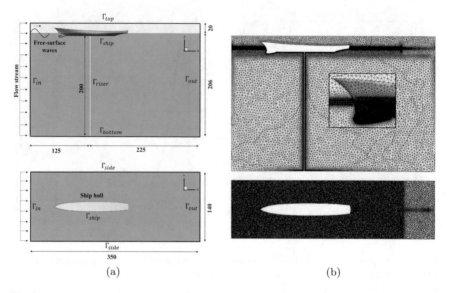

Fig. 8 Demonstration of coupled drillship-riser system: **a** schematic of the full-scale DTMB 5415 with the computational domain and **b** computational mesh (the inset provides the discretization of the ship hull)

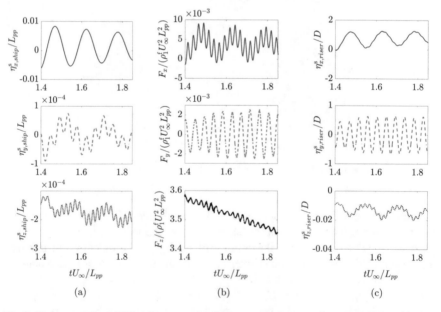

Fig. 9 Flow across the drillship-riser system: **a** displacement response (η^s_{ship}/L_{pp}) of the drillship, **b** fluid forces ($F/(\rho^f_1 U^2_\infty L^2_{pp})$) on the drillship, and **c** displacement response (η^s_{riser}/D) at the midpoint of the riser of diameter D at $z/L = 0.5$; in the surge (X), sway (Y), and heave (Z) directions

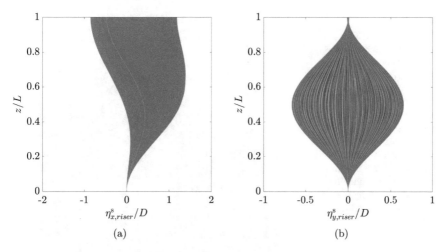

Fig. 10 Response envelope along the span of the riser in **a** in-line (X) and **b** cross-flow (Y) directions

The displacement and hydrodynamic force responses for the drillship are shown in Fig. 9a, b, respectively, in the surge (X), sway (Y), and heave (Z) directions. The displacement response of the riser mid-point at $z/L = 0.5$ is shown in Fig. 9c, corresponding to the location with the maximum amplitude response along the riser. The cross-flow response of the drillship manifests two frequencies with $fL_{pp}/U_\infty = 5.2$ being the dominant one. It is observed that the dominant in-line and cross-flow frequencies along the riser are in synchronization with the surge and sway frequencies of the drillship. The riser oscillates with a high amplitude ($\sim \mathcal{O}(D)$) albeit its reduced velocity being away from the typical "lock-in" range of $U_r \in [4-6]$. This high amplitude may be attributed to the synchronization of the riser response amplitude with that of the vessel, leading to vessel-induced motion of the riser. The response envelope of the riser depicting the amplitude response along the riser with temporal variation is shown in Fig. 10 where we observe a standing wave-like pattern in the cross-flow response.

The contour plot of the vessel and the riser at $tU_\infty/L_{pp} = 1.535$ is shown in Fig. 11 where we observe the formation of the Kelvin waves downstream of the vessel and the vortex shedding process along the riser. Note that the computational mesh employed for the demonstration is coarse. Further analyses employing a finer mesh is required for detailed physical understanding.

6 Conclusions

The present work discusses the novel formulation for the numerical modeling of two-phase turbulent flows interacting with multibody structures. The formulation benefits

Fig. 11 Coupled drillship-riser system subjected to free-surface waves and ocean current at $tU_\infty/L_{pp} = 1.535$. The free surface of the waves is represented by the iso-surface of the order parameter at $\phi = 0$ colored by the elevation. The inset provides the Z-vorticity contours along the riser surface with red and blue colors representing positive and negative vorticities, respectively

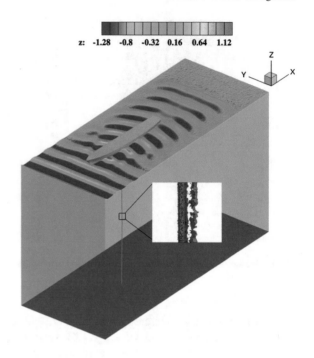

from the advantages of both exact tracking of the fluid-structure interface by ALE technique, and diffused interface capturing of the two-phase fluid-fluid interface via phase-field Allen-Cahn equation. Furthermore, the partitioned strategy for the coupling allows for flexibility and ease in numerical implementation of the formulation. The solver has been extensively validated with standard benchmark problems. Applications involving complex internal two-phase flow in pipelines subjected to flow externally have been carried out. Finally, the solver has been demonstrated for a drillship-riser system where the drillship interacts with the free-surface ocean waves and the riser is exposed to turbulent ocean current. The results show a coupled response between the drillship and the riser, which is very imperative to study for predicting the behavior of such complex system in drastic ocean environments.

References

1. Sorensen RM (1993) Basic wave mechanics: for coastal and ocean engineers. A Wiley-Interscience Publication, Wiley
2. Chakrabarti SK (1987) Hydrodynamics of offshore structures. Springer
3. Blevins RD (1990) Flow-induced vibration. Van Nostrand Reinhold, New York
4. Paidoussis MP, Price SJ, de Langre E (2010) Fluid-Structure interactions: cross-flow-induced instabilities. Cambridge University Press
5. Mittal R, Iaccarino G (2005) Immersed boundary methods. Ann Rev Fluid Mech 37(1):239–261

6. Sotiropoulos F, Yang X (2014) Immersed boundary methods for simulating fluid-structure interaction. Prog Aerosp Sci 65:1–21
7. Donea J, Giuliani S, Halleux JP (1982) An arbitrary Lagrangian-Eulerian finite element method for transient dynamic fluid-structure interactions. Comput Methods Appl Mech Eng 33:689–723
8. Unverdi SO, Tryggvason G (1992) A front-tracking method for viscous, incompressible, multi-fluid flows. J Comput Phys 100(1):25–37
9. Vignjevic R, Campbell J (2009) Review of development of the smooth particle hydrodynamics (SPH) method. Springer US, Boston, MA, pp 367–396
10. Sethian JA, Smereka P (2003) Level set methods for fluid interfaces. Ann Rev Fluid Mech 35:341. Copyright—Copyright Annual Reviews, Inc. 2003; Last updated—18 May 2014
11. Scardovelli R, Zaleski S (1999) Direct numerical simulation of free-surface and interfacial flow. Ann Rev Fluid Mech 31(1):567–603
12. Anderson DM, McFadden GB, Wheeler AA (1998) Diffuse-interface methods in fluid mechanics. Ann Rev Fluid Mech 30(1):139–165
13. Hron J, Turek S, A monolithic FEM/multigrid solver for an ALE formulation of fluid-structure interaction with applications in biomechanics. In: Hans-Joachim B, Michael S (eds), Fluid-Structure interaction, Berlin, Heidelberg, 2006. Springer, Berlin, Heidelberg, pp 146–170
14. Bazilevs Y, Calo VM, Hughes TJR, Zhang Y (2008) Isogeometric fluid-structure interaction: theory, algorithms, and computations. Comput Mech 43(1):3–37
15. Gee MW, Küttler U, Wall WA (2010) Truly monolithic algebraic multigrid for fluid-structure interaction. Int J Numer Methods Eng 85(8):987–1016
16. Felippa CA, Park KC, Farhat C (2001) Partitioned analysis of coupled mechanical systems. Comput Methods Appl Mech Eng 190(24):3247–3270. Advances in Computational Methods for Fluid-Structure Interaction
17. Piperno S, Farhat C, Larrouturou B (2001) Partitioned procedures for the transient solution of coupled aeroelastic problems—part II: energy transfer analysis and three-dimensional applications. Comput Methods Appl Mech Eng 190:3147–3170
18. Jaiman R, Geubelle P, Loth E, Jiao X (2011) Combined interface boundary condition method for unsteady fluid-structure interaction. Comput Methods Appl Mech Eng 200(1):27–39
19. Jaiman R, Geubelle P, Loth E, Jiao X (2011) Transient fluid-structure interaction with non-matching spatial and temporal discretizations. Comput Fluids 50(1):120–135
20. Jaiman RK, Pillalamarri NR, Guan MZ (2016) A stable second-order partitioned iterative scheme for freely vibrating low-mass bluff bodies in a uniform flow. Comput Methods Appl Mech Eng 301:187–215
21. Förster C, Wall WA, Ramm E (2007) Artificial added mass instabilities in sequential staggered coupling of nonlinear structures and incompressible viscous flows. Comput Methods Appl Mech Eng 196:1278–1293
22. Dettmer WG, Peric D (2007) A fully implicit computational strategy for strongly coupled fluid-solid interaction. Archiv Comput Methods Eng 14:205–247
23. Jaiman RK, Guan MZ, Miyanawala TP (2016) Partitioned iterative and dynamic subgrid-scale methods for freely vibrating square-section structures at subcritical Reynolds number. Comput Fluids 133:68–89
24. Joshi V, Jaiman RK (2019) A hybrid variational Allen-Cahn/ALE scheme for the coupled analysis of two-phase fluid-structure interaction. Int J Numer Methods Eng 117(4):405–429
25. Joshi V, Jaiman RK (2018) A positivity preserving and conservative variational scheme for phase-field modeling of two-phase flows. J Comput Phys 360:137–166
26. Allen SM, Cahn JW (1979) A microscopic theory for antiphase boundary motion and its application to antiphase domain coarsening. Acta Metallurgica 27(6):1085–1095
27. Kim J, Lee S, Choi Y (2014) A conservative Allen-Cahn equation with a space-time dependent Lagrange multiplier. Int J Eng Sci 84:11–17
28. Joshi V, Jaiman RK (2017) A positivity preserving variational method for multi-dimensional convection-diffusion-reaction equation. J Comput Phys 339:247–284

29. Joshi V, Jaiman RK (2018) An adaptive variational procedure for the conservative and positivity preserving Allen-Cahn phase-field model. J Comput Phys 366:478–504
30. Brackbill JU, Kothe DB, Zemach C (1992) A continuum method for modeling surface tension. J Comput Phys 100(2):335–354
31. Kim J (2012) Phase-field models for multi-component fluid flows. Commun Comput Phys 12(3):613–661, 009
32. Kim J (2005) A continuous surface tension force formulation for diffuse-interface models. J Comput Phys 204(2):784–804
33. Spalart PR, Allmaras SR (1994) A one-equation turbulence model for aerodynamic flows. La Rech. Aérospatiale 1:5–21
34. Joshi V, Jaiman RK (2017) A variationally bounded scheme for delayed detached eddy simulation: application to vortex-induced vibration of offshore riser. Comput Fluids 157:84–111
35. Li Y, Law YZ, Joshi V, Jaiman RK (2018) A 3D common-refinement method for non-matching meshes in partitioned variational fluid–structure analysis. J Comput Phys 374:163–187
36. Joshi V, Jaiman RK, Ollivier-Gooch C (2020) A variational flexible multibody formulation for partitioned fluid-structure interaction: application to bat-inspired drones and unmanned air-vehicles. Comput Math Appl 80(12):2707–2737
37. Gurugubelli PS, Ghoshal R, Joshi V, Jaiman RK (2018) A variational projection scheme for nonmatching surface-to-line coupling between 3D flexible multibody system and incompressible turbulent flow. Comput Fluids 165:160–172
38. Saad Y, Schultz MH (1986) GMRES: a generalized minimal residual algorithm for solving nonsymmetric linear systems. SIAM J Sci Stat Comput 7(3):856–869
39. MPI (2015) A Message-Passing Interface Standard, Version 3.1. Technical report
40. Mao X, Joshi V, Jaiman R (2021) A variational interface-preserving and conservative phase-field method for the surface tension effect in two-phase flows. J Comput Phys 433:110166
41. Nogueira E, Cotta RM (1990) Heat transfer solutions in laminar co-current flow of immiscible liquids. Wärme - und Stoffübertragung 25(6):361–367
42. Kajitani H, Miyata H, Ikehata M, Tanaka H, Adachi H, Namimatsu M, Ogiwara S (1983) The summary of the cooperative experiment on Wigley parabolic model in Japan. In: 17th ITTC resistance committee report
43. US Navy Combatant, DTMB 5415. http://www.simman2008.dk/5415/combatant.html

Multiscale Modeling of Chromatin Considering the State and Shape of Molecules

Yuichi Togashi

Abstract Chromatin is a complex of DNA and proteins in the eukaryotic cell nucleus. This highly complex structure helps not only to fold long DNA and compactly store it into the nucleus but also to control how the information on the DNA is processed. Hence, the structural dynamics of chromatin is important for the understanding of genomic processes. To track the behavior, as well as imaging experiments, molecular dynamics simulations are adopted. However, as the chromatin in the whole nucleus is huge, it is always required to reduce the computational cost to a realistic level. Hence, coarse-grained and multiscale models have been developed, although it is still underway. In this Chapter, recent modeling efforts for the chromatin structure, including our adaptive resolution approach and future directions, are briefly introduced.

Keywords DNA · Nucleosome · Chromatin · Molecular machine ·
Coarse-graining · Multiscale model · Molecular dynamics simulation

1 Introduction

Genetic information is coded on genomic DNA and stored in the nucleus in eukaryotic cells. For example, each human cell has 46 chromosomes including 2 m of DNA in total. To pack them into the 10-μm-sized nucleus, DNA is folded with help of binding proteins, and the DNA–protein complex is called chromatin. Typically,

Y. Togashi (✉)
College of Life Sciences, Ritsumeikan University, 1-1-1 Noji-higashi, Kusatsu, Shiga 525-8577, Japan
e-mail: togashi@fc.ritsumei.ac.jp

RIKEN Center for Biosystems Dynamics Research, 3-10-23 Kagamiyama, Higashi-Hiroshima, Hiroshima 739-0046, Japan

Research Center for the Mathematics on Chromatin Live Dynamics, Hiroshima University, 1-3-1 Kagamiyama, Higashi-Hiroshima, Hiroshima 739-8526, Japan

© The Author(s), under exclusive license to Springer Nature Singapore Pte Ltd. 2022   171
R. Srinivas et al. (eds.), *Advances in Computational Modeling and Simulation*,
Lecture Notes in Mechanical Engineering,
https://doi.org/10.1007/978-981-16-7857-8_14

DNA is wrapped around histone proteins to form barrel-like structures called nucleo-somes, which then stack one another to form chromatin fibers, and are finally shaped into chromosomes. On the other hand, to read out the DNA information, molecular machines such as replicases must access the DNA strand, which may be blocked if the DNA is tightly packed with proteins. That is even utilized for regulation of gene expression; there are regions called *heterochromatin*, where DNA is so tightly packed with proteins that the expression of genes there is strongly suppressed, in contrast to *euchromatin* permissible for gene expression. In other words, the *shape*, or how the DNA is folded, can determine how the information recorded there is read and used. In this sense, DNA is not merely a data tape but an information-processing machine by itself. Hence, the structure and dynamics of chromatin have attracted attention in recent years. Experimental observation techniques such as single-molecule imaging and next-generation sequencers enabled us to track the structural changes. Still, it is difficult to track the conformational changes as a whole and over time experimen-tally, and hence modeling and simulation of chromatin dynamics are expected to complement experimental data and are currently under development.

2 Modeling and Simulation of Chromatin

2.1 Molecular Dynamics Simulation

Molecular dynamics simulation is a popular tool for tracking conformational change and motion of molecules or molecular complexes. It is based on classical (Newto-nian) dynamics. When we use an all-atom model, typically, each atom is modeled as a material point, and the force on each atom is calculated according to a potential func-tion (called force field) depending on the conformation i.e. positions of the atoms. By simply solving the equation of motion numerically, we can track the motion of each atom. There are popular force fields such as AMBER or CHARMM, which have been improved over years and used for a variety of biomolecules. However, the compu-tational cost is so large that the applicable timescale is still limited to only micro- to milli-seconds. Hence, simplified and coarse-grained models are wanted, particu-larly for large molecular complexes such as chromatin. As a bottom-up approach, there are coarse-grained models in which a group of atoms (e.g. an amino acid residue in a protein) is represented by a point, e.g. MARTINI and Gō models. In case further simplification is required, more abstract models e.g. bead-spring polymers or even continuum models (e.g. phase-field model [1]) are used. In such drastically simplified models, unlike all-atom models, it is difficult to construct the force-field or governing equation from the first principles. For chromatin, all-atom or residue-level coarse-grained models have been applied at the microscopic nucleosome level, and polymer models or reaction–diffusion-like models have been adopted for the macro-scopic chromosome or nucleus level; however modeling scheme at the intermediate mesoscopic level has not been fully established yet.

2.2 Multiscale Modeling

Apart from the case of chromatin, there is a methodology called multiscale modeling, to deal with systems where both microscopic properties and macroscopic behavior should be considered. Multiscale models are typically constructed in such a bottom-up way; first, a small part of the system is simulated at the most microscopic scale using a detailed (often first-principle) model, and properties or phenomenological laws are extracted from the result to construct a model at the next scale, i.e., a slightly coarse-grained and/or simplified model. This procedure can be iteratively applied to generate a model at the next scale, to obtain a hierarchy of models. Finally, these models are simulated concertedly. For example, starting from the molecular level, a multiscale model was constructed for a whole human heart, which could hence reproduce e.g. how mutation of the protein affects the dynamics of the heart [2]. Of course, the model at the microscopic scale can be used only for a small part and a short time; otherwise, the computational cost will be unrealistic. Fortunately, it usually works well because, in general, 1. Within a short time, the effects of an event can propagate only within a small region, and 2. At the macroscopic scale, in the long run, fast behavior and fluctuations are averaged out, and only the mean properties and their slow changes govern the system; thus, it is enough to consider only fast microscopic behavior and slow macroscopic behavior.

2.3 Adaptive Resolution Approach

In some cases, there are regions where microscopic details are particularly important for the process. If the important regions are fixed throughout the simulation, we can construct a hybrid model combining a detailed high-resolution model for such important regions and a simplified low-resolution model for the others; methods for multi-scale modeling would be useful for properly connecting the regions represented by different models. It is however not always guaranteed. For example, when we consider the searching process by DNA-binding proteins for certain binding sites on DNA, the DNA segments near the protein molecules are important for the process, which should of course move as the proteins travel. Ideally, we can efficiently simulate such a system if the resolution of each region in the hybrid model can be adjusted on the fly (Fig. 1), and that is possible if the models are reversibly transferred to each other by the following conversions: 1. a model to a lower resolution model (coarse-graining); 2. a model to a higher resolution model ("fine-graining"). The former is usually simple and just similar to the multi-scale modeling case. The latter is however complicated, as the microscopic information is missing in the original (lower resolution) model and hence must be inferred or reconstructed without discrepancy.

We tried to construct such an adaptive resolution model and succeeded for a simple Rouse polymer case [3]. The Rouse model consists of material points connected by linear springs. To increase the resolution of a certain part, points must be added, as

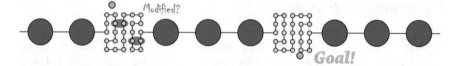

Fig. 1 Schematic representation of the concept of the adaptive resolution approach. Only important regions, e.g. near the walker molecule or with special modification, are simulated at a high resolution; otherwise at a low resolution

well as the parameters for the points and springs should be adjusted. In this simple case, such points can be relatively easily sampled, not violating the equilibrium distribution, and the parameter conversion is quite simple. In contrast, points are removed when the resolution is decreased. Note that, information on the microscopic conformation represented by those points is lost upon the removal; when the resolution is increased again, the positions of the new points are resampled regardless of the previous conformation.

3   Current Problems and Outlook

3.1   Difficulties Intrinsic to Chromatin

As mentioned above, we constructed an adaptive resolution model for Rouse polymers. Based on this, we aimed to include features of chromatin fibers, to attain a comprehensive and efficient modeling framework for the structural dynamics of chromatin. Nevertheless, there is a pitfall intrinsic to the case of chromatin.

Suppose that we reduce the resolution for a part of the model and then increase the resolution back again. As already mentioned, the system forgets the previous microscopic conformation when the resolution is reduced, and the conformation is resampled when the resolution is recovered. For the resampling, we generally assume a (quasi-)steady-state distribution depending only on the variables (conformation) of the lower resolution model, although not necessarily in equilibrium in terms of physics, and the new conformation to be sampled is independent of the previous one. As microscopic structures rapidly fluctuate, we can assume this for simple polymers as long as the interval between these changes of the resolution is long enough.

However, in the case of chromatin, there are microscopic modifications such as DNA methylation or histone modification. Although such a modification itself is a fast process typically exerted by enzymatic reactions, the intervals of modification events can be very long (e.g. days). Hence, it may work as a long-term memory, which is a unique feature of chromatin and important for its biological function. At the same time, though, microscopic behavior cannot be averaged out even at the timescale of the macroscopic model, which means that the methodology of the adaptive resolution modeling and also typical multiscale modeling breaks down. Particularly, when the lifetime of the memory is longer than typical switching intervals of the model resolution, the precondition for the independent resampling of conformation is no longer valid.

3.2 Future Directions

This problem arises from the fundamental nature of chromatin and is hence difficult to solve. One way to mitigate the problem is by combining the model with experiments. Recently, we proposed a method to estimate the fractal dimension and size of chromatin domains from the sub-diffusional motion of single nucleosomes there [4]. Fractal dimension indicates how compactly the chromatin fiber is folded, which may be useful in the *fine-graining* operation in the adaptive resolution method, to generate more realistic microscopic structures from a small number of parameters. Also, we suggested a bead-spring polymer-like model for chromatin, for which parameters of the springs are estimated from a contact map obtained by the Hi-C experiment [5]. Using this method, we can simulate dynamics of chromatin corresponding to a certain experimental dataset or even reproduce motion connecting (morphing) multiple contact maps. Complementary to such data-driven approaches, artificial or abstract models may be useful for realizing concepts, e.g. the *state* and operation cycles of molecular machines could be incorporated into reaction–diffusion models [6] and bead-spring polymer models.

Recently, as experimental techniques have been evolving, huge data sets obtained by different methodologies have been accumulated in databases. An important role of modeling and simulation is to combine such multi-modal data into a comprehensive model and show the underlying mechanism in a comprehensible manner. We hope that the abovementioned approaches including ours will be further improved, to faithfully play this role.

Acknowledgements YT's works were in part supported by JSPS KAKENHI JP18KK0388.

References

1. Seirin-Lee S, et al (2019) PLoS Comput Biol 15, e1007289
2. Sugiura S et al (2012) Prog Biophys Mol Biol 110:380–389
3. Rolls E et al (2017) Multiscale Model Simul 15:1672–1693
4. Shinkai S, et al (2016) PLoS Comput Biol 12, e1005136
5. Shinkai S, et al (2020) NAR Genom Bioinfo 2, lqaa020
6. Togashi Y (2019) J Phys Chem B 123:1481–1490

Numerical Methods for the Isoperimetric Problem on Surfaces

Amit R. Singh

Abstract The isoperimetric problem on a surface is to find a sub-surface that has a specified area and the least possible perimeter. We discuss the development of a numerical technique to identify locally minimizing sub-surface for a given surface and area. The numerical technique is applied to some sample surfaces for varying prescribed areas and the results are presented.

1 Introduction

The isoperimetric problem has an interesting mythological connection. Ancient Roman and Greek records mention the story of Dido, the queen of Carthage, a city in ancient Phoenicia. She fled the tyrannical rule of her brother and founded the city of Carthage in North Africa. But there she had to strike a deal with a local monarch who put forth an interesting proposition to Dido—she could own as much land as she could surround with the hide of a bull. Dido decided to cut the bull's hide in narrow strips and enclosed a vast circular land out of it! Thus, the intelligent Dido solved the extremal geometry problem of what is the maximum area one can enclose by a given length of the perimeter. This is the classical isoperimetric problem on a plane surface. The solution, as Dido found, is a circle. There are many analogs and variations of this problem. For example, the inverse problem is to identify the surface of the smallest area that can span a specified closed space curve. This is called the Plateau's problem. In high-school physics, we learn that the sphere encloses the maximum volume with the least surface area. This is a three-dimensional extension of the same problem. The isoperimetric problem also arises in geometry-driven flow. For instance, it is known that a geometric flow of a surface curve driven by its geodesic curvature under the constraint of a fixed enclosed surface area necessarily minimizes the length of the closed curve [5]. Thus, the isoperimetric problem on a three-dimensional surface physically represents a two-phase flow where one of the

A. R. Singh (✉)
Department of Mechanical Engineering, BITS Pilani, Pilani, India
e-mail: amit.singh@pilani.bits-pilani.ac.in

© The Author(s), under exclusive license to Springer Nature Singapore Pte Ltd. 2022   177
R. Srinivas et al. (eds.), *Advances in Computational Modeling and Simulation*,
Lecture Notes in Mechanical Engineering,
https://doi.org/10.1007/978-981-16-7857-8_15

phases forms an island in the second phase and it flows under the line tension at the interface of the two phases.

In the following sections, we will develop some numerical tools step by step to find approximate solutions to the isoperimetric problem on three-dimensional surfaces. We will restrict our discussions to Monge patches i.e. surfaces which can be expressed as z = f(x, y) [3]. We will choose a sample surface for which the minimizing solution can be determined by visual inspection. We will validate our numerical methods for this surface. First, we will show a direct approach to minimize the perimeter under a global area constraint. We will try to understand the numerical artifacts of the direct approach. Next, we will discuss some strategies to overcome the limitations of the direct approach.

2 Direct Method

Given a 3D surface in the Monge form and an area value less than the total area of the surface, we have to find the location of a sub-surface that has the lowest perimeter. To achieve this, we will start with an initial guess for the sub-surface. We will discretize the sub-surface into a mesh of 3D triangles. The x- and y-coordinates of the vertices of the triangles will be the unknown degrees of freedom of our optimization problem. We will identify the edges of the triangles that lie on the boundary of the meshed region. The sum of the lengths of the boundary edges will be the objective function for our optimization problem. We will include a constraint on the sum of the areas of all the triangles as a penalty term in the objective function. We will minimize the objective function using the L-BFGS-B [10] routine.

2.1 Discretization of the Initial Guess

Intuitively, we can say that the perimeter enclosed by a surface area will be the minimum if the surface is a geodesic disk. If the surface is a flat plane, this gives us a flat circular disk as the minimizer. So we will start with an initial guess where the projection of the Monge patch in the $(x$-$y)$-plane is a circular disk. Of course, this does not imply that the 3D surface will be a geodesic disk but it is a reasonable initial guess for the optimization routine. A good quality mesh is desirable for numerical techniques.

To discretize the 2D circular projection, we will make use of the following algorithm to first generate vertices of the circular mesh. The vertices will be connected into triangles using the Delaunay triangulation algorithm [2]. The advantage of Algorithm 1 is that it allows us to have triangles of specified size almost uniformly distributed over the mesh. The desired element size is an input to the algorithm.

In Algorithm 1, we calculate the radius of the circular disk from the given area and then divide the radius into equal parts based on the input element size. The number of

---

Algorithm 1 Generate vertices for triangulation of a circular disk.

1: **function** CREATEVERTICES(A, x, y, s)   ▷ $A \rightarrow$ area, $(x, y) \rightarrow$ center, $s \rightarrow$ element size
2:     $R \leftarrow \sqrt{A/\pi}$ ▷ $R \rightarrow$ radius of the disk
3:     $N_{\text{rings}} \leftarrow \text{round}(R/s)$
4:     $N_{\text{points}} \leftarrow \text{round}\left(3N_{\text{rings}}\left(N_{\text{rings}}+1\right)+1\right)$
5:     Points $\leftarrow$ Empty array of N_{points} vertices
6:     $i \leftarrow 1$ ▷ Counter for the number of vertices generated
7:     **for** $j \leftarrow 1, N_{\text{rings}}$ **do**
8:         $r \leftarrow j\left(R/N_{\text{points}}\right)$ ▷ Radius of the jth ring
9:         $N_{\text{ringpoints}} \leftarrow 6j$ ▷ Number of points in the jth ring
10:        $\Delta\theta \leftarrow 2\pi/N_{\text{ringpoints}}$
11:        **for** $k \leftarrow 1, N_{\text{ringpoints}}$ **do**
12:            $\theta \leftarrow (k-1)\Delta\theta$
13:            Points[i] $\leftarrow \begin{Bmatrix} x + r\cos\theta \\ y + r\sin\theta \end{Bmatrix}$
14:            $i \leftarrow i + 1$
15:        **end for**
16:    **end for**
17:    Points[i] $\leftarrow \begin{Bmatrix} x \\ y \end{Bmatrix}$ ▷ Add a vertex at the center
18:    **return** Points
19: **end function**

---

parts the radius is divided into gives us the number of "rings" in the mesh. Then we place points at equal angular spacing along each ring. Figure 1a shows an example of the ring of points obtained using this algorithm. Figure 1b shows the Delaunay triangulation of the vertices of Fig. 1a. The edges in red in Fig. 1b are the boundary edges which are identified by the fact that they belong to only one triangle in the mesh unlike the interior edges (in gray) which are shared between two triangles.

To obtain the initial guess for the optimization problem, we will project the vertices of the triangulation of the circular disk obtained using Algorithm 1 to the 3D surface. This requires that the equation of the Monge patch $z = f(x, y)$ is known to us. For example, if the Monge patch equation is (surface rendered in Fig. 2)

$$z = \left(10e^{\frac{5x}{2}+\frac{5y}{2}} - 9e^{\frac{5y}{2}} + 8 - 7e^{\frac{5x}{2}}\right)e^{-\frac{x^2}{8}-\frac{5x}{4}-\frac{y^2}{8}-\frac{5y}{4}-\frac{25}{4}}, \tag{1}$$

after projecting the z-coordinates of the vertices of the mesh of Fig. 1b we get the 3D initial guess for our optimization problem as shown in Fig. 3.

2.2 Calculating the Perimeter Length and the Surface Area

The optimization algorithm requires that we are able to evaluate the objective function for a given set of optimization degrees of freedom values. Our objective function is

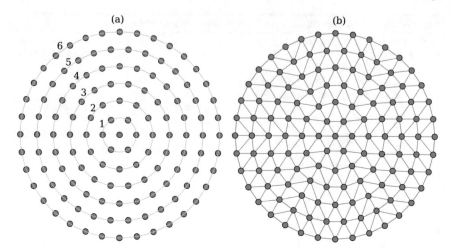

Fig. 1 **a** Vertices generated using Algorithm 1 for a circular disk of area $A = 9\pi$ and element size 0.5. The total radius of the disk is $R = 3$ and we get six rings shown as gray circles. **b** Delaunay triangulation of vertices of Fig. 1a. The red edges mark the boundary edges which occur only in one triangle

Fig. 2 The 3D surface of Eq. 1

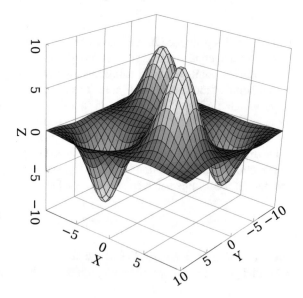

Fig. 3 The 3D mesh obtained using Algorithm 1 after projecting the vertices of Fig. 1b to the surface given by Eq. 1. The center of the 2D projected disk is at (2.5, 5.0). The actual surface area of the mesh is 63.85 although the area of the mesh projected on the (x-y)-plane is 9π

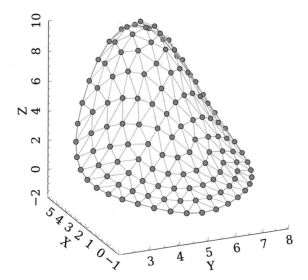

$$\min \mathcal{F}(\mathbf{X}) = \mathcal{L}(\mathbf{X}) + \sum_{\text{triangle}} \frac{1}{2} k \left(\mathcal{A}\left(\mathbf{X}_{\text{triangle}}\right) - A_{0,\text{triangle}} \right)^2, \qquad (2)$$

where $\mathcal{L}(\mathbf{X})$ is the length of the perimeter as a function of the degrees of freedom $\mathbf{X}$ and $\mathcal{A}\left(\mathbf{X}_{\text{triangle}}\right)$ is the surface area of a triangle of the 3D mesh as a function of the degrees of freedom $\mathbf{X}$ (the x, y-coordinates of the vertices of the mesh). We have imposed the area constraint $\mathcal{A}\left(\mathbf{X}_\triangle\right) = A_{0,\text{triangle}}$ as a penalty term with coefficient k in Eq. 2 for every triangle. The more $\mathcal{A}\left(\mathbf{X}_{\text{triangle}}\right)$ deviates from the constrained value, the objective function increases quadratically. Thus, the optimization algorithm will try to keep the penalty terms small.

The advantage of the direct method lies in the simplicity of calculating $\mathcal{L}(\mathbf{X})$ and $\mathcal{A}\left(\mathbf{X}_{\text{triangle}}\right)$. We have already identified the boundary edges (shown as red in Fig. 1b and Fig. 3) using the criterion that each boundary edge occurs only in one triangle in the mesh. Therefore,

$$\mathcal{L}(\mathbf{X}) = \sum_{\substack{\text{boundary} \\ \text{edges}}} L_{\text{edge}} \qquad (3)$$

$$L_{\text{edge}}\left(\mathbf{X}_i, \mathbf{X}_j\right) = \|\mathbf{X}_i - \mathbf{X}_j\|, \qquad (4)$$

where L_{edge} is the Euclidean distance between the vertices with position vectors $\mathbf{X}_i$ and $\mathbf{X}_j$ that constitute the edge. The total surface area of the meshed sub-surface can be calculated as

$$\sum_{\text{triangle}} \mathcal{A}\left(\mathbf{X}_{\text{triangle}}\right) = \sum_{\text{triangles}} A_{\text{triangle}}\left(\mathbf{X}_i, \mathbf{X}_j, \mathbf{X}_k\right) \tag{5}$$

$$A_{\text{triangle}}\left(\mathbf{X}_i, \mathbf{X}_j, \mathbf{X}_k\right) = \frac{1}{2}\|\left(\mathbf{X}_j - \mathbf{X}_i\right) \times \left(\mathbf{X}_k - \mathbf{X}_i\right)\| \tag{6}$$

with A_{triangle} as the area of a triangular element of the mesh made up of the vertices with position vectors $\mathbf{X}_i$, $\mathbf{X}_j$, and $\mathbf{X}_k$.

2.3   Calculating the Derivative of the Objective Function

The L-BFGS-B optimization routine that we plan to use requires an analytical derivative of the objective function with respect to the degrees of freedom. It is possible to use numerical differentiation to approximate the derivative but it will slow down the optimization program a lot. We can calculate it as

$$\frac{\mathcal{F}}{\mathbf{X}} = \frac{\mathcal{L}}{\mathbf{X}} + \sum_{\text{triangle}} k\left(\mathcal{A}\left(\mathbf{X}_{\text{triangle}}\right) - A_{0,\text{triangle}}\right)\frac{\mathcal{A}}{\mathbf{X}_{\text{triangle}}}. \tag{7}$$

In Eq. 7, the derivatives are vectors of the same shape as $\mathbf{X}$. To proceed further, we note that to calculate the derivatives of $\mathcal{L}$ and $\mathcal{A}$, we only need to be able to calculate the derivatives of L_{edge} and A_{triangle}. Now we will make use of index notation [7] to make the analytical calculations tractable.

Let X_{ai} denote the coordinates of the ith vertex where $a = 1, 2$ indicate the x- and y-coordinates of the vertex, respectively. We can write length of an edge made up of the ith and the jth vertices as

$$L_{\text{edge}}\left(\mathbf{X}_i, \mathbf{X}_j\right) = \left[\sum_{a=1}^{2}\left(X_{ai} - X_{aj}\right)\left(X_{ai} - X_{aj}\right)\right]^{1/2}, \tag{8}$$

where the repeated index a indicates summation which has also been explicitly shown. But we will avoid showing the summation explicitly from now on. Next, let us differentiate Eq. 8 with respect to the coordinates of the kth vertex.

$$\frac{L_{\text{edge}}}{X_{bk}} = \left[\left(X_{ai} - X_{aj}\right)\left(X_{ai} - X_{aj}\right)\right]^{-1/2}\left(X_{ai} - X_{aj}\right)\left(\frac{X_{ai}}{X_{bk}} - \frac{X_{aj}}{X_{bk}}\right) \tag{9}$$

$$= \frac{1}{L_{\text{edge}}}\left(X_{ai} - X_{aj}\right)\left(\delta_{ab}\delta_{ik} - \delta_{ab}\delta_{jk}\right),$$

where δ is the Kronecker Delta [8].

We can write the area of a triangle in index notation as

$$A_{\text{triangle}}\left(\mathbf{X}_i, \mathbf{X}_j, \mathbf{X}_k\right) = \frac{1}{2}\left[(X_{aj} - X_{ai})(X_{bk} - X_{bi})\,\epsilon_{abc}\,(X_{dj} - X_{di})(X_{ek} - X_{ei})\,\epsilon_{dec}\right]^{1/2},$$
$$\tag{10}$$

where ϵ_{abc} denotes the Levi-Civita [9] symbol. Using the property of Levi-Civita symbol, we can write

$$\epsilon_{abc}\epsilon_{dec} = \delta_{ad}\delta_{be} - \delta_{ae}\delta_{db}.$$

This simplifies Eq. 10 as

$$A_{\text{triangle}}\left(\mathbf{X}_i, \mathbf{X}_j, \mathbf{X}_k\right) = \frac{1}{2}\left[(X_{aj} - X_{ai})^2\,(X_{bk} - X_{bi})^2 - (X_{aj} - X_{ai})(X_{bk} - X_{bi})(X_{bj} - X_{bi})(X_{ak} - X_{ai})\right]^{1/2}.$$
$$\tag{11}$$

Now we can write the derivative as

$$\frac{\partial A_{\text{triangle}}}{\partial X_{cl}} = \frac{1}{8A_{\text{triangle}}}\Big(-2\left(X_{ai} - X_{aj}\right)^2\left(X_{ci} - X_{cl}\right) + \left(X_{ai} - X_{aj}\right)\left(X_{ai} - X_{al}\right)\left(X_{ci} - X_{cj}\right)$$
$$+ \left(X_{ai} - X_{ak}\right)\left(X_{ai} - X_{al}\right)\left(X_{ci} - X_{ck}\right) + 2\left(-X_{aj} + X_{al}\right)^2\left(-X_{ck} + X_{cl}\right)$$
$$- \left(-X_{aj} + X_{al}\right)\left(-X_{ak} + X_{al}\right)\left(-X_{cj} + X_{cl}\right) - \left(-X_{aj} + X_{al}\right)\left(-X_{ak} + X_{al}\right)\left(-X_{ck} + X_{cl}\right)$$
$$+ \left(X_{bi} - X_{bj}\right)\left(X_{bi} - X_{bl}\right)\left(X_{ci} - X_{cj}\right) - 2\left(X_{bi} - X_{bk}\right)^2\left(X_{ci} - X_{cl}\right)$$
$$+ \left(X_{bi} - X_{bk}\right)\left(X_{bi} - X_{bl}\right)\left(X_{ci} - X_{ck}\right) - \left(-X_{bj} + X_{bl}\right)\left(-X_{bk} + X_{bl}\right)\left(-X_{cj} + X_{cl}\right)$$
$$- \left(-X_{bj} + X_{bl}\right)\left(-X_{bk} + X_{bl}\right)\left(-X_{ck} + X_{cl}\right) + 2\left(-X_{bk} + X_{bl}\right)^2\left(-X_{cj} + X_{cl}\right)\Big).$$
$$\tag{12}$$

With the equations for the objective function and its derivative with respect to the degrees of freedom, we have all the information needed to solve the optimization problem.

3   Numerical Simulations

We will solve our optimization problem on a sample surface for some constrained area values and different initial guesses.

3.1   Shortcomings of the Direct Method

Let us use the surface of Eq. 1 as the underlying surface and the mesh of Fig. 3 as the initial guess. This implies that the constrained area is 9π and our initial guess is the sub-surface centered at $(2.5, 5.0)$. The result of optimizing Eq. 2 with the initial guess as the mesh shown in Fig. 3 is shown in Fig. 4. The key observation is that in the final state the mesh is all jumbled up and it has not reached the expected position on top of the "peak" on which it is located.

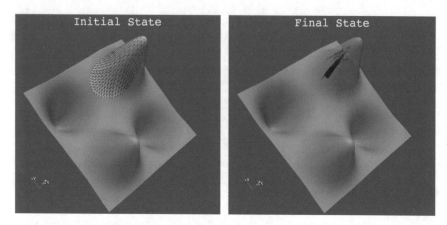

Fig. 4 On the left: the initial guess mesh of Fig. 3 overlayed on top of the surface of Eq. 1. **On the right**: The mesh obtained as a result of the optimization algorithm applied to the initial guess

Why has the solution mesh jumbled up? The answer lies in the nature of the objective function. The area term in Eq. 2 is defined as a summation over each triangle of the mesh. Certainly, the vertices of the mesh in the interior are shared between multiple triangles. So a change in the coordinates of the vertex due to the area term will affect multiple triangles. Yet, there is nothing in the objective function that ensures that the vertices should displace in a way that maintains continuity of the displacement field across the triangles. In other words, although we have expressed the objective function in terms of the triangles, after substituting the coordinates of the vertices, the information about the juxtaposition of the triangular elements is retained only in the terms of coincident vertices. The triangles are thus free to fold over themselves or even overlap or cross over each other as the vertices are iteratively displaced during the optimization routine. Mathematically, the final state shown in Fig. 4 minimizes the sum of the lengths of the boundary edges while retaining the constrained value of the areas of the triangles. So it is a mathematical minimizer but because it lacks the non-overlapping feature of the triangles of the initial mesh it is not the physically desirable solution.

3.2 Improving the Direct Method Solution Strategy

As we have noted in the previous section, we need to incorporate information about spatial continuity across the triangles in our numerical scheme to avoid the non-physical solution shown in the final state of Fig. 4. To achieve this end, we will borrow some tricks from the Finite Element Method [11].

1. We already have a triangular mesh. We will equip them with linear triangular shape functions (refer [11] or any standard textbook on the Finite Element Method for more details on linear triangular elements).
2. We will redefine our objective function in terms of a displacement field defined over the initial mesh instead of solving directly for the final coordinates of the solution mesh. The mathematical formulation of the basis functions of the linear triangular finite elements ensures continuity of the displacement field across the elements and this should address our issue of the triangles crossing over each other during the optimization iterations.

$$\min \mathcal{F}\left(\mathbf{u}\right) = \mathcal{L}\left(\mathbf{X}+\mathbf{u}\right) + \sum_{\text{triangle}} \frac{1}{2} k \left(\mathcal{A}\left(\mathbf{X}_{\text{triangle}} + \mathbf{u}_{\text{triangle}}\right) - A_{0,\text{triangle}}\right)^2 \quad (13)$$

$$\mathbf{u}_a = \sum_{\text{vertex } i}^{\text{all vertices}} u_{ai} N_i(\xi, \eta) \qquad \text{where } a = 1, 2. \quad (14)$$

Here, $\mathbf{u}_a$ denotes the x, y-components of the displacement field of the vertices of the mesh. $N_i(\xi, \eta)$ are the shape functions of the "standard isoparametric" linear triangular finite element (see [11] or any other standard text on the finite element method for the meaning of "isoparametric" and "standard" elements). Figure 5 shows the "standard" linear triangle element. The equations for the length of the boundary and its derivative with respect to the components of the displacement degrees of freedom can be written analogously to Eqs. 8 and 9. We discuss the calculations for the areas next.

3. Instead of calculating areas of the triangles in terms of the vertex coordinates directly, we will make use of differential geometry and redefine area of a triangle as an integration over the triangle's projection in the $x - y$-plane. The integration will be converted into a summation using Gaussian quadrature [1] that is commonly used in finite element analysis.

   The differential area of a Monge patch $z(x, y)$ is given as

Fig. 5 The standard linear triangle with second-order Gauss quadrature points shown as the red dots

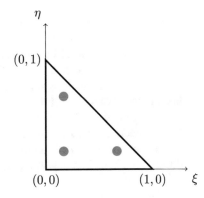

$$ds^2 = \sqrt{EG - F^2} dxdy \quad \text{where} \tag{15}$$

$$E = \left(1 + \frac{\partial z}{\partial x}^2\right) \tag{16}$$

$$F = \frac{\partial z}{\partial x} \frac{\partial z}{\partial y} \tag{17}$$

$$G = \left(1 + \frac{\partial z}{\partial y}^2\right). \tag{18}$$

The area of a 3D triangle of the mesh can then be written as

$$A_{\text{triangle}} = \int_\Delta \sqrt{EG - F^2}\, dxdy. \tag{19}$$

A common mathematical convenience used in the finite element method is to map the $(x\text{-}y)$-coordinates of the triangles to a "standard reference triangle" shown in Fig. 5. This leads to simplification in carrying out numerical integration over the triangles as the shape function and its derivatives can be calculated just once for the reference triangle and reused for the remaining triangles. The mapping between the physical and the reference coordinates is captured through the Jacobian J defined in Eq. 20. Then, we can rewrite the area integral of Eq. 19 as Eq. 21 where $|J| = \det J$.

$$J = \begin{bmatrix} \frac{\partial x}{\partial \xi} & \frac{\partial x}{\partial \eta} \\ \frac{\partial y}{\partial \xi} & \frac{\partial y}{\partial \eta} \end{bmatrix} \tag{20}$$

$$A_{\text{triangle}} = \int_0^1 \int_0^{1-\xi} \sqrt{EG - F^2}\, |J| d\eta d\xi \tag{21}$$

$$= \sum_{q=1}^3 w_q \sqrt{EG - F^2}\, |J|. \tag{22}$$

In Eq. 22, we convert the integration over the standard triangle as a summation using Gauss quadrature of second order. The three Gauss quadrature points are shown in Fig. 5. It should be noted that in Eq. 21, the integrand must be evaluated at (x, y) values corresponding to (ξ, η) values at the Gauss quadrature points.

4. After all the modifications discussed above, our objective function will be a function of unknown u_{ai} which are the displacements of the vertices of the mesh in the x- and y-directions. To carry out optimization, we need to calculate the derivative of the objective function with respect to u_{ai}. This is non-trivial and to avoid errors it is recommended to use automatic differentiation [4]. We can also use symbolic differentiation and code generation abilities of packages like SymPy [6].

5. To gain insight into the trajectory of the degrees of freedom as the optimization iterations proceed, we will use the gradient descent method for solving our optimization problem. Gradient descent can be written as

$$\mathbf{u}^{n+1} = \mathbf{u}^n - \gamma \frac{\partial \mathcal{F}}{\partial \mathbf{u}^n}. \tag{23}$$

Mathematically, we move the degrees of freedom in the direction of the steepest descent with a step size controlled by γ. The termination criterion can be when

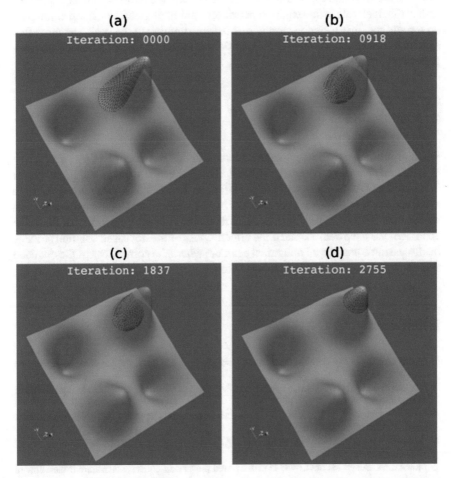

Fig. 6 The results of the simulations after implementing the improvements to the direct method as discussed in Sect. 3.2. The iteration number "0000" corresponds to the degrees of freedom after the first 1000 iterations. The other iteration numbers should be multiplied by 1000. The key result is that the mesh has converged on the peak of the surface which offers the least perimeter for the given area constraint

the change $\|\mathbf{u}^{n+1} - \mathbf{u}^n\| <$ tolerance is satisfied for several iterations which may be accompanied with the norm of the gradient $\|\frac{\partial \mathcal{F}}{\partial \mathbf{u}^n}\|$ remaining very small for several iterations.

3.3   Improved Results

The simulation results for the direct method with the improvements discussed in Sect. 3.2 are shown in Fig. 6. As seen from the figure, the mesh does not get jumbled up and the final state is the expected perimeter minimizing state at the top of the peak on the surface. It is also noteworthy that before climbing to the peak, the mesh first shrinks to bridge the gap between the initial area value of 63.85 and the constrained area value of $9\pi \approx 28.37$.

4   Conclusion

We have presented a direct numerical method to find the location of a surface patch of specified area value lying on a general 3D surface $z(x, y)$ such that its perimeter is minimum. We have discussed the challenges one can encounter when taking a naive approach to solve the problem. We have also presented some numerical techniques borrowed from the finite element method to mitigate the issues of the naive approach. We have demonstrated the effectiveness of the improved solution strategy by discussing a non-trivial numerical example where the correctness of the obtained solution can be verified by visual inspection.

References

1. Cowper GR (1973) Gaussian quadrature formulas for triangles. Int J Numer Methods Eng 7(3):405–408
2. Delaunay B, et al (1934) Sur la sphere vide. Izv Akad Nauk SSSR, Otdelenie Matematicheskii i Estestvennyka Nauk 7(793–800):1–2
3. Gray A, Abbena E, Salamon S (1997) Modern Differential Geometry of Curves and Surfaces with Mathematica. CRC Press, 2nd edn
4. Griewank A, et al (1989) On automatic differentiation. Mathematical programming: recent developments and applications 6(6):83–107
5. KolÁŘ M, BeneŠ M, ŠevČoviČ D (2017) Area preserving geodesic curvature driven flow of closed curves on a surface. Discret Contin Dyn Syst-B 22(10):3671–3689
6. Meurer A, Smith CP, Paprocki M, Čertík O, Kirpichev SB, Rocklin M, Kumar A, Ivanov S, Moore JK, Singh S, Rathnayake T, Vig S, Granger BE, Muller RP, Bonazzi F, Gupta H, Vats S, Johansson F, Pedregosa F, Curry MJ, Terrel AR, Roučka Š, Saboo A, Fernando I, Kulal S, Cimrman R, Scopatz A (2017) Sympy: symbolic computing in python. Peer J Comput Sci 3:e103

7. Stover C, Weisstein EW (1999) Einstein summation. Last accessed on 02 Aug 2021. https://mathworld.wolfram.com/EinsteinSummation.html
8. Weisstein EW (1999) Kronecker delta. Last accessed on 02 Aug 2021. https://mathworld.wolfram.com/KroneckerDelta.html
9. Weisstein EW (1999) Permutation symbol. Last accessed on 02 Aug 2021. https://mathworld.wolfram.com/PermutationSymbol.html
10. Zhu C, Byrd RH, Lu P, Nocedal J (1997) Algorithm 778: L-bfgs-b: Fortran subroutines for large-scale bound-constrained optimization. ACM Trans Math Softw (TOMS) 23(4):550–560
11. Zienkiewicz OC, Taylor RL, Nithiarasu P, Zhu JZ (1977) The finite element method, vol 3. McGraw-hill London

QSAR—An Important In-Silico Tool in Drug Design and Discovery

Ravichandran Veerasamy

Abstract QSAR (Quantitative structure–activity relationship) study is important thirst area in drug design and discovery via computational studies of chemistry. The hypothesis, alterations in structure of molecules reflect proportional variations in the pharmacological or biological activity, is the centre for focus of QSAR analysis. Currently one dimensional to six dimensional QSAR methods is available, and they are used in lead optimization, classification, and prediction of pharmacological or biological activity, pharmacokinetic properties, and toxicity of chemical compounds. The accomplishment of various models of QSAR depends on many factors or criteria's such as input data accuracy, selection of descriptors, feature selection, model development, and validation as reported in OECD principle for QSAR. Validation is an important step in QSAR, and it is used to establish reliability and significance of a procedure of specific purpose. However, to be useful, QSAR models should be revealing and easily understandable in explaining the essential molecular features that plays a major role in the alteration of biological activity. Thus, the goal of the current chapter is to briefly discuss the principle, prerequisites to set up correct models, methods, validation, limitations of model applications, and the significance of QSAR in drug discovery.

Keywords Quantitative structure–activity relationship · Chemo-informatics · Randomization · Applicability domain · Applications in drug discovery

R. Veerasamy (✉)
Pharmaceutical Chemistry Unit, Faculty of Pharmacy, AIMST University, 08100 Semeling, Kedah, Malaysia
e-mail: ravichandran_v@aimst.edu.my

Centre of Excellence for Biomaterial Engineering, AIMST University, 08100 Semeling, Kedah, Malaysia

1 Introduction

Drug design and development is an essential research area in pharmaceutical industry, and it has many challenges and hurdle such as problem in efficacy and delivery, long time span, and cost effect. One of the main obstacles with the study was the complex and big data that were received from different genomic studies and clinical trials. All the above-mentioned issues and challenges in drug discovery are solved by artificial intelligence (AI) and machine learning (ML) technology, and these algorithms have modernized the drug designing. ML algorithms are already used in numerous drug discovery methods such as peptide synthesis, virtual screening (structure and ligand based), drug monitoring and release, pharmacophore modelling, QSAR, drug repositioning, poly-pharmacology, toxicity, and physiochemical properties forecasting [1]. As total, computer aided drug design (CADD) is one of the important areas in the current arena of drug development process (Fig. 1).

In silico approaches which are based on the chemistry-biology-informatics triad can bring pharmacology to new heights. Over the last 50 years, QSAR models have played a major role in establishing relationship between chemical's molecular structure and their biological effects like potency, toxicity, ADME, and physico-chemical properties [2]. Numerical representatives of 2D and 3D molecular descriptors of chemical structures are a vital feature of QSAR. QSAR concept has also been used in virtual high throughput screening (VHTS), and it becomes an essential part of lead discovery process.

In the process of drug discovery, QSAR has begun and progressed to fulfil the needs and desire of chemists. Kubinyi (2002) summarized the detailed history of QSAR in one of his inspiring articles [3]. QSAR is a ligand-based approach (Fig. 2) since it uses the information of ligands in model development. The ligand-based approaches are most appropriate method in drug design when there is a lack of information about the various targets. The motives for using QSAR and quantitative structure–property relationship (QSPR) models in drug discovery are (i) to cut down time duration and cost; (ii) to make robust calculation of activities/properties; (iii) to avoid unwanted synthesis; (iv) to extract details from big data; (vi) to gather details about MoA of biological activities. Both approaches have broad applications in life sciences (biology, agriculture and medicine) [4], as well as physical sciences

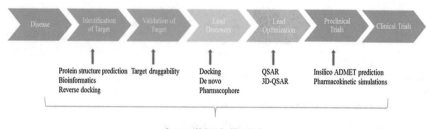

Fig. 1 Role of computer aided drug design in drug discovery

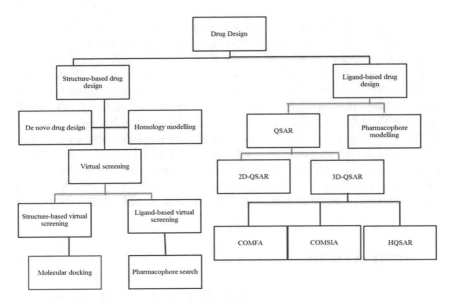

Fig. 2 Different in-silico tools used in drug design

(organic chemistry, physical chemistry, materials sciences) [5]. It is certainly a tough task to design a drug for the interested target while it shows proper pharmacokinetic properties and be devoid of toxicity. Gleeson (2008) discovered a set of easy and explainable and interpretable rules to recognize the source of ADMET properties by simple forms like molecular weight, partition coefficient (log$P$) ionization state, etc. [6].

The robustness of QSAR is mainly altered by the prediction ability of that model. Principally, all the compounds in a dataset are described by a set of quantitative or qualitative independent variables that are mathematically related with the biological (pIC$_{50}$) or chemical (i.e. log$P$) endpoint of interest by using statistical analysis. However, the QSAR models are most suitable for the data that was used to develop it. QSAR models should be deliberated as a complementary tool for assisting the decision-making process, but not pondered as the replacement for knowledge of the scientist [7].

Another advantage of QSAR is to design or analyse ligands for multiple/dual targets using the additivity of the molecular fields [8, 9]. For example, Huang et al. (2010) used 3D-QSAR techniques like CoMFA and CoMSIA models to find out the multitarget activity of PPAR agonist to treat obesity and stroke [8]. The major drawbacks of QSAR include false correlations due to the experimental error in biological data, and many QSAR results are failure to produce or predict the particular activity. Generally, QSAR studies contain various challenges like effect of quality and quantity of underlying training data, incompatible calculations by different QSAR models, predictive power of QSAR tools varies with the chemical set, limitation in knowing the knowledge of the drug action in the whole body, and insufficient description of some important interactions. QSAR is used in finding out the mechanism of action of drug, identify and analyse the pharmacological or biological activity as well as toxicity, optimization of lead, environmental chemistry, and nano chemistry.

The complete utilization of QSAR models has not yet been attained because the present research is mainly focussed on creating good predictive models but failing in interpretability. Most robust ML algorithms are used in producing predictive QSAR models with improved predictive ability, but they are not providing any information about the specific or underlying features that influencing the activity, and so-called as a *black box*. Further, a careful data curation is essential for a robust and reliable model. This can only be achieved by expertise of trained practitioners, since all the data are not modellable or may fail to provide accurate results all time since there are several inherent issues. Most of those problems have been deliberated in this chapter.

2 History of QSAR

Actually, the foundation of QSAR (molecular structures directly influence the biological activity) was begun when Cros (1863) observed an inverse correlation among toxicity and aqueous solubility [10]. In 1962, Hansch et al. officially introduced the term QSAR [11]. Free and Wilson in 1964, introduced a simple and effective method of QSAR, which was named as 'Free-Wilson model' [12]. The Free-Wilson model is a numerical method which correlates structural topographies (presence or absence of functional groups) with biological properties. However, both Hansch and Free-Wilson methods are closely interrelated in theoretical view along with their practical applicability [13].

Later mixed approach (Hansch and Free-Wilson) has been used to explain the relation between the interested biological activity and both physicochemical and Free-Wilson type parameters [13, 14]. Fruitful applications of this mixed model approach on the SAR of enzyme inhibitors [15–22], revealed the better performance of mixed classical QSAR. Later, the non-classical QSAR approach—model developed from a huge, heterogeneous, and data with numerous MoA—was introduced [23]. Then the QSAR models for numerous compounds against various target proteins (so-called proteochemometric, also named as computational chemogenomics) was ensued [24, 25]. In 2006, Kubinyi have reported the success stories of QSAR with other CADD methods [26].

3 QSAR Methodology

As mentioned earlier, classical to six dimensional QSAR methods are available [27]. Conformational arrangement of atoms in space does not affect 2D-QSAR, while 3D-QSAR needs the 3D structures of ligand and require alignment of ligands in three spatial dimensions. 4D-QSAR represents multiple conformations, orientations, and

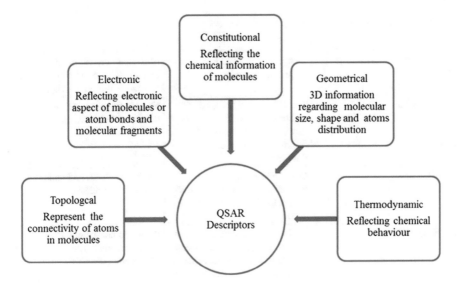

Fig. 3 Descriptors used in classical QSAR studies

the protonation state of ligand [28–32]. Induced-fit scenarios of ligands on active site represents 5D and solvation models can be thought of 6D-QSAR, respectively. 7D is real target-based receptor model. Molecular descriptors (Fig. 3, Table 1) used in QSAR model as independent variables are numerical representative of chemical data of a molecular structure.

The general QSAR model can be expressed as:

$$\text{Predicted Biological Activity} = \text{Function (Chemical Structure)}$$

The QSAR model development workflow can be split into three main steps such as preparation of data analysis of data, and model validation (Fig. 4). Acquiring a predictive QSAR model was influenced by some of the factors such as the of endpoint data and descriptors quality, feature selection, validation of various methods used to develop various kind of models [33–35]. In the first decade of twenty-first century, the European Organization for Economic Co-operation and Development (OECD) has formed a set of standards—appropriate measures of goodness-of-fit, robustness, and predictive potential—for the development and validation of QSAR models [36]. The OECD guidance especially emphasizes on validation of developed QSAR model by using sets of compounds that were not actually included in the model development.

Table 1 Different dimensional molecular descriptors

Descriptor	Description	Example
0D QSAR	Descriptors derived from molecular formula	Molecular weight, number and type of atoms, polarizability, volume, electronegativity, etc.
1D QSAR	The 1D descriptors are counts and properties of the functional groups and sub-structural fragments, also referred as fingerprints	Functional groups, rings, bonds, substituents, etc.
2D QSAR	Obtained from graph theoretical representation such as topological and connectivity descriptors	Total path count, molecular connectivity indices, etc.
3D QSAR	Encode various geometrical as well as 3D spatial information of a molecule in terms of their shape, steric and electronic features	Polar and non-polar surface area, molecular volume, and other geometrical properties
4D QSAR	The fourth dimension is an ensemble of conformation of each ligand	—
5D-QSAR	Like 4D QSAR with added advantage of explicit demonstration of diverse induced-fit models	—
6D-QSAR	Further incorporating different solvation scenarios in 5D-QSAR	—
Physicochemical properties	These are calculated measured quantities based on parameterization with measured data	$\log P$, pKa, solubility measures, etc.

Table 2 Summary of OECD principles for QSAR modelling

No	OECD Principles	Description
1	Defined endpoint	To ensure that all endpoint values within a given data set are consistent
2	Unambiguous algorithm	To ensure transparency and reproducibility of the proposed QSAR model
3	Defined applicability domain	To determine the boundaries in which the model is robust for predicting query compounds
4	Measures of model's predictive potential	To evaluate the internal and external predictive power of the model
5	Mechanistic interpretation	To ensure that the underlying mechanism of action of compounds can be elucidated

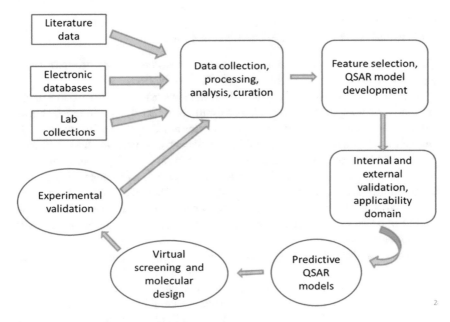

Fig. 4 Workflow of QSAR model development

3.1 Statistical Methods Used in QSAR Modelling

3.1.1 Linear Methods (LM)

Simple linear regression (SLR), multiple linear regression (MLR), principal component regression (PCR), partial least squares (PLS) regression, genetic function approximation (GFA), genetic partial least squares (G/PLS) techniques, etc. are the various LM used in QSAR modelling.

PLS regression method is exclusively useful when the number of descriptors is more than the number of compounds and/or there is a correlation between variables [37, 38]. MLR is a mathematical technique used to predict some unknown dependent variables from several independent variables [39]. K-means clustering is top-down approach, and in this method the objects are segregated into a fixed number (k) of cluster based on the similarity in variables or descriptor value [40]. PCA is used to estimate the number of clusters in the data [41]. Genetic algorithm (GA) is used in finding solutions for optimization. The general application of GA is topology optimizations, genetic training algorithm, and control parameters optimization.

3.1.2 Nonlinear Methods (NM)

- NM can be obtained from more complicated methodologies

- LM becomes NM if nonlinear terms of some descriptors are added.
- Many NM are developed from LM via kernel trick.

Nonlinear methods include support vector machine (SVM), support vector regression (SVR), artificial neural network (ANN or NN), etc.

ANN is an interconnection group of artificial neurons that uses a mathematical model or computational model. For example, in 2019, Zuvela et al. showed that the antioxidant activity of flavonoids is determined by using this method. The two well-established antioxidant activity mechanisms, namely, the hydrogen atom transfer (HAT) mechanism defined with the minimum bond dissociation enthalpy, and the sequential proton-loss electron transfer (SPLET) mechanism defined with proton affinity and electron transfer enthalpy [42]. Zhou et al. (2015) have developed nonlinear QSAR models with high-dimensional descriptor selection and SVR to improve toxicity prediction and evaluation of phenols on *Photobacterium phosphoreum* [43].

SVM are a group of related supervised learning methods that can be used for the classification and regression. A special property of SVM is that they simultaneously minimize the empirical classification error and geometric margin. Ramandi et al. (2020) have constructed QSAR models by combining Genetic Algorithms with Multiple Linear Regressions (GA-MLR) and Support Vector Machine (SVM) to explore the anticoagulant activity of factor Xa inhibitors [44]. Decision forest (DF) is a decision support tool which can be applied to regression and classification models. Random forest (RF) is an ensemble of unpruned regression tree [45, 46] that is used to construct the model with 2/3 of a training data set and the dataset to evaluate the model predictive power.

3.2 Validation

The usage of a QSAR model is mainly depending on the calculation capacity of model. For good prediction, the QSAR model must be properly validated, otherwise it leads to false prediction. So, validation is a vital step in QSAR studies [35]. Only in the last two decades, validation of QSAR models has acquired significant attention [33–35, 47, 48]. The validity of a QSAR model is assessed by (i) cross/internal validation, (ii) bootstrapping, (iii) randomization, and (iv) external validation. Numerous principles were framed for judging the validity of QSAR models at an international workshop held in Setubal (Portugal), which were subsequently modified in 2004 by the OECD Work Programme on QSARs [36, 49]. The formula for calculation of different internal and external validation parameters and threshold values are given in Tables 3 and 4, respectively. The external validation technique is always considered as a best validation method in QSAR modelling.

Bootstrap resampling (or bootstrapping) (Efron 1979) is one more technique used for internal validation—alternative to LMO [50]. A high average q^2 in the bootstrap validation confirm the model robustness. Further, Y-randomization test is used to

Table 3 Formula for calculation of some important QSAR model validation parameters

Validation parameters

$$q^2 = 1 - \frac{\sum \left(y_i - \hat{y}_i\right)^2}{\sum \left(y_i - y_{mean}\right)^2}$$

where y_i, and $\hat{y}_i$ are the actual and predicted activity of the ith molecule in the training set, respectively, and y_{mean} is the average activity of all molecules in the data set

$$pred\_r^2 = 1 - \frac{\sum \left(y_i - \hat{y}_i\right)^2}{\sum \left(y_i - y_{mean}\right)^2}$$

where y_i, and $\hat{y}_i$ are the actual and predicted activity of the ith molecule in the test set, respectively, and y_{mean} is the average activity of all molecules in the data set

$$Q^2_{F1} = 1 - \frac{\sum \left(y_i - \hat{y}_i\right)^2}{\sum \left(y_i - \bar{y}_{TR}\right)^2}$$

$$Q^2_{F2} = 1 - \frac{\sum \left(y_i - \hat{y}_i\right)^2}{\sum \left(y_i - \bar{y}_{EXT}\right)^2}$$

$$Q^2_{F3} = 1 - \frac{\left[\sum \left(y_i - \hat{y}_i\right)^2\right]/n_{EXT}}{\sum \left(y_i - \bar{y}_{TR}\right)^2/n_{TR}}$$

$$r_m^2 = r^2 \left(1 - \sqrt{|r^2 - r_0^2|}\right)$$

where r^2 is the squared correlation coefficient between observed and predicted values and r_0^2 is the squared correlation coefficient between observed and predicted values with intercept value set to zero of test set

$$r_m'^2 = r^2 \left(1 - \sqrt{|r^2 - r_0'^2|}\right)$$

where r^2 is the squared correlation coefficient between observed and predicted values and $r_0'^2$ is the squared correlation coefficient between predicted and observed values with intercept value set to zero of test set

$$\overline{r_m^2} = \frac{r_m^2 + r_m'^2}{2}$$

$$\Delta r_m^2 = \left|r_m^2 - r_m'^2\right|$$

$$r_p^2 = r^2 \left(1 - \sqrt{|r^2 - r_r^2|}\right)$$

where r_r^2 is squared mean correlation coefficient of the randomized models and r^2 is squared correlation coefficient of the non-randomized model

$$CCC = \frac{2\sum (y_i - \bar{y})\left(\hat{y}_i - \overline{\hat{y}}\right)}{\sum (y_i - \bar{y})^2 + \sum \left(\hat{y}_i - \overline{\hat{y}}\right)^2 + n_{EXT}\left(\bar{y} - \overline{\hat{y}}\right)^2}$$

where y is experimental data and $\hat{y}$ is external prediction data

Table 4 Minimum criteria for some common statistical parameters in QSAR validation

Statistical parameters	Validation	Conditions*
r^2	Internal	>0.6 (for in vivo data)
CCC_{tr}	Internal	>0.85
q^2_{LOO}	Internal	>0.5
CCC_{cv}	Internal	>0.85
q^2_{LMO}	Internal	>0.5
pred_r^2	External	>0.6
CCC_{ext}	External	>0.85
q_2-F_1	External	>0.6**
q_2-F_2	External	>0.6**
q_2-F_3	External	>0.6**
r^2_m	External	>0.5***
r'^2_m	External	>0.5
Δr^2_m	External	< 0.2
r^2_m average	External	> 0.5
k'	External	0.85 < k' < 1.15
k	External	0.85 < k < 1.15
r^2-r^2_0/r^2	External	<0.1
r^2-r'^2_0/r^2	External	<0.1
r^2_m(overall)	Internal and External	>0.5
r'^2_m(overall)	Internal and External	>0.5
r^2_m average (overall)	Internal and External	>0.5
Δr^2_m (overall)	Internal and External	<0.2
r^2_p	Internal (Randomization)	>0.5

* The criteria may change depending on the QSAR model development method used and the heterogeneity of scaffold. ** The criteria may be > 0.70 and *** > 0.65 as proposed by Chirico et al. (n.d.) [53]

confirm model robustness. In Y-randomization, the dependent variable values are scrambled, and all the calculations can be repeated for at least five (better, more) times. The main of this method is to confirm that the obtained models' good statistics are not due to over-fitting [51, 52]. These statistical methodologies are used to confirm the sound and unbiasedness ("good model") of generated QSAR models.

A designed model can be agreed generally in QSAR (MLR, PLS regression, etc.) studies when it fulfils (but not always) the limit values given in Table 4 (the values may be the minimum recommended values for significant QSAR model). The other threshold values to accept the QSAR models are 1. The standard deviation 's' is not greater than standard deviation of the biological data. 2. The F value indicates that overall significance level (better than 95%). 3. The confidence interval of all individual regression coefficients proves that they are justified at the 95% significance level. 4. Randomized r^2 value should be as low as to r^2. 5. Randomized q^2 value should be as low as to q^2. Moreover, the biological data should be well distributed with at least two or even more logarithmic units. Also, physicochemical parameter should be scaled and with free from collinearity issues [52].

3.2.1   Dispute on QSAR Validation

Hawkins et al. [54, 55] have claimed that it is good to better to utilize LOO cross-validation when the sample size is small, since it is meaningless to divide the data set in to training and test [54, 55]. An inconsistency among the internal and external prediction was stated in a few QSAR studies [56, 57], and some other studies claimed that there is no association between internal and external prediction [58]. In many cases, it is very difficult to find out the final topmost model because some models had better internal validation parameters than external one, and vice-versa. Other than the mentioned validation parameters, the real outliers and structural similarity must be analysed, and necessary action must be implemented to rectify the issues.

4   Nano—QSAR Model

The basic principle of QSAR for nanoparticles or nano-QSAR is focused on calculating the mathematical relationship between the variance in molecular properties programmed, hence known as nano-descriptors and the variance in biological activity for a set of nanomaterial.

5   Errors in QSARs/QSPRs

The various kind of errors obtained during QSAR/QSPR development are depicted in Table 5, along with the respective OECD principle(s) [59].

Outlier is a data point which has high standardized residuals when compared with other samples of the data set (biological space) or the compound fall outside the applicability domain (chemical space) (Fig. 5). In Fig. 5, compound number 283 is an activity outlier since the compound standardized residual value is more than the recommended + 2.5 standardized residual, and 347 is a structural and activity outlier since the compound is far away from the recommended standardized residual of -2.5 and the calculated Hat value of 2.45. Both the compounds 283 and 347 are belong to prediction set. The biological space outlier can be carefully eliminated from QSAR

Table 5 Common types of error found in QSAR/QSPR model development

No	OECD Principles	Type of error
1	Defined endpoint	Failure to consider data heterogeneity account Inclusion of undefined biological values
2	Unambiguous algorithm	Use of collinear descriptors Use of incomprehensible descriptors Error in descriptor values
3	Defined applicability domain	Inadequate/undefined applicability domain Use of inadequate data Duplication of compounds in dataset Unacknowledged omission of data points Too narrow range of endpoint values
4	Measures of model's predictive potential	Over-fitting of data Lack of descriptor auto-scaling Inadequate training/test set selection Lack of/inadequate/ misuse of statistics Use of excessive numbers of descriptors in QSAR Failure to consider distribution of residuals Failure to do internal or external or both internal and external validation
5	Mechanistic interpretation	Lack of mechanistic interpretation

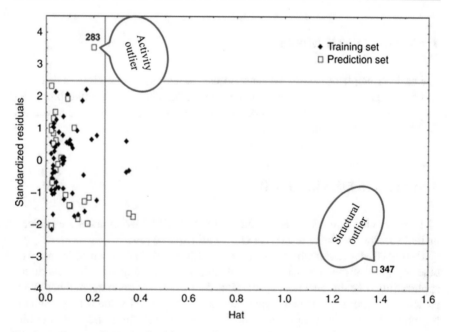

Fig. 5 Williamson plot for applicability domain

Table 6 Applicability domain techniques

Technique	Method
Ranges in descriptor space	Bounding box PCA bounding box
Distance-based	Leverage approach Euclidean distance Mahalanobis distance KNN approach Tanimoto similarity
Probability density distribution	Parametric method Non-parametric method
Response variable based	Range of the experimental data
Miscellaneous	Standardization approach Kernel-based

models. However, chemical space outlier needs further attention. The following are the reasons why some of the compounds are outlier in a congeneric data set: 1. may have different mechanism; 2. may interact with its respective target in different mode; 3. conformational flexibility of target protein binding site [60–62]. The applicability domain (AD) of a QSAR method describes the model limitations with relates to its structural subspace and response space [63]. The different techniques used to find out AD are given in Table 6 [64].

6 Open-Source Software for QSAR Modelling

Some open-source software's utilised for calculation of descriptors and quantitative structure activity modelling are given in Table 7.

7 Conclusion

The accomplishment of drug discovery mainly based on the utilisation of various kind of SAR techniques. QSAR study was initially developed by Corwin Hansch and his colleagues about more than 55 years ago. Over the years, there is a lot of growth in QSAR field such as various new types of algorithm and techniques. In spite of certain drawbacks, the QSAR concept remains as one of the active areas in drug design development and discovery. The disadvantages and faults in QSAR are continuously being recognised, resolved, and improved by researchers to develop a more robust QSAR model. The OECD standards and principles in QSAR modelling will help the research community to develop good-fit, reliable, robust, and predictive QSAR models in drug design and discovery.

Table 7 Resources and software for performing QSAR modelling

	Software/Web server	References
Open-source descriptor calculation software	CDK	[65, 66]
	PaDEL	[67, 68]
	RCPI	[69]
	Chemical Descriptors Library (CDL)	[70]
	ChemoPy	[71]
	Babel	[72]
	MORDRED	[73]
	RDkit	[74]
Software for performing QSAR modelling	AutoWeka	[75]
	AZOrange	[76]
	CDK-Taverna	[77]
	CHARMMing	[78]
	ChemBench	[79]
	ChemMine	[80]
	CORAL	[81]
	DMax Chemistry Assistant	[82]
	OCHEM	[83]
	OCED QSAR Toolbox	[84]
	PASS Online	[85]
	QSARINS	[86]
	QSAR Workbench	[87]
	Toxtree	[88]
	LQTAgrid	[89]
	QSAR-Co-X	[90]
	DTC-QSAR	[91]
	DPubChem	[92]
	BioPPSy	[93]

References

1. Gupta R, Srivastava D, Sahu M, Tiwari S, Ambasta RK, Kumar P (2021) Artificial intelligence to deep learning: machine intelligence approach for drug discovery. Mol Diversity 25:1315–1360
2. Khan MTH, Sylte I (2007) Predictive QSAR modeling for the successful predictions of the ADMET properties of candidate drug molecules. Curr Drug Discov Technol 4:141–149
3. Kubinyi H (2002) From narcosis to hyperspace: The history of QSAR. Quant Struct-act Relatsh 21:348–356

4. Prachayasittikul V, Worachartcheewan A, Shoombuatong W, Songtawee N, Simeon S, Prachayasittikul V, Nantasenamat C (2015) Computer-aided drug design of bioactive natural products. Curr Top Med Chem 15:1780–1800

5. Katritzky AR, Kuanar M, Slavov S, Hall CD, Karelson M, Kahn I, Dobchev DA (2010) Quantitative correlation of physical and chemical properties with chemical structure: utility for prediction. Chem Rev 110:5714–5789

6. Gleeson MP (2008) Generation of a set of simple, interpretable ADMET rules of thumb. J Med Chem 51:817–834

7. Shoombuatong W, Prathipati P, Owasirikul W, Worachartcheewan A, Simeon S, Anuwongcharoen N, Wikberg JES, Nantasenamat C (2017) Towards the revival of interpretable QSAR models. Challenges and advances in computational chemistry and physics. Springer International Publishing, Cham, pp 3–55

8. Huang HJ, Lee KJ, Yu HW, Chen HY, Tsai FJ, Chen CY (2010) A novel strategy for designing the selective PPAR agonist by the "sum of activity" model. J Biomol Struct Dyn 28(2):187–200

9. Sundriyal S, Bharatam PV (2009) 'Sum of activities' as dependent parameter: a new CoMFA-based approach for the design of pan PPAR agonists. Eur J Med Chem 44(1):42–53

10. Cros AFA (1863) Action de lalcohol amylique sur lorganisme. PhD thesis, University of Strasbourg

11. Hansch C, Maloney PP, Fujita T, Muir RM (1962) Correlation of biological activity of phenoxyacetic acids with Hammett substituent constants and partition coefficients. Nature 194:178–180

12. Free SM Jr, Wilson JW (1964) A mathematical contribution to structure-activity studies. J Med Chem 7:395–399

13. Kubinyi H (1988) Free Wilson analysis. Theory, applications and its relationship to Hansch analysis. Quant Struct-act Relatsh 7:121–133

14. Wei DB, Zhang AQ, Han SK, Wang LS (2001) Joint QSAR analysis using the Free-Wilson approach and quantum chemical parameters. SAR QSAR Environ Res 12:471–479

15. Verma RP, Hansch C (2009) Camptothecins: a SAR/QSAR study. Chem Rev 109:213–235

16. Gao H, Katzenellenbogen JA, Garg R, Hansch C (1999) Comparative QSAR analysis of estrogen receptor ligands. Chem Rev 99:723–744

17. Selassie CD, Garg R, Kapur S, Kurup A, Verma RP, Mekapati SB, Hansch C (2002) Comparative QSAR and the radical toxicity of various functional groups. Chem Rev 102:2585–2605

18. Kurup A, Garg R, Carini DJ, Hansch C (2001) Comparative QSAR: angiotensin II antagonists. Chem Rev 101:2727–2750

19. Hansch C, Gao H (1997) Comparative QSAR: radical reactions of benzene derivatives in chemistry and biology. Chem Rev 97:2995–3060

20. Hansch C, Hoekman D, Gao H (1996) Comparative QSAR: toward a deeper understanding of chemicobiological interactions. Chem Rev 96:1045–1076

21. Hadjipavlou-Litina D, Garg R, Hansch C (2004) Comparative quantitative structure—activity relationship studies (QSAR) on non-benzodiazepine compounds binding to benzodiazepine receptor (BzR). Chem Rev 104:3751–3794

22. Garg R, Kurup A, Mekapati SB, Hansch C (2003) Cyclooxygenase (COX) inhibitors: a comparative QSAR study. Chem Rev 103:703–732

23. Fujita T, Winkler DA (2016) Understanding the roles of the "two QSARs." J Chem Inf Model 56:269–274

24. Cortés-Ciriano I, Ain QU, Subramanian V et al (2015) Polypharmacology modelling using proteochemometrics (PCM): recent methodological developments, applications to target families, and future prospects. Medchemcomm 6:24–50

25. Qiu T, Qiu J, Feng J, Wu D, Yang Y, Tang K, Cao Z, Zhu R (2017) The recent progress in proteochemometric modelling: focusing on target descriptors, cross-term descriptors and application scope. Brief Bioinform 18:125–136

26. Kubinyi H (2006) Success stories of computer-aided design. Computer applications in pharmaceutical research and development. Wiley, Hoboken, NJ, USA, pp 377–424

27. Cronin MTD, Livingstone DJ (2004) Predicting Chemical Toxicity and Fate. CRC Press LLC, Boca Raton, FL, London
28. Albuquerque MG, Hopfinger AJ, Barreiro EJ, Alencastro RB (1998) Four-dimensional quantitative structure—activity relationship analysis of a series of interphenylene 7-oxabicycloheptane oxazole thromboxane A2 receptor antagonists. J Chem Inf Comput Sci 38:925–938
29. Santos-Filho OA, Hopfinger AJ (2001) A search for sources of drug resistance by the 4D-QSAR analysis of a set of antimalarial dihydrofolate reductase inhibitors. J Comput Aided Mol Des 15:1–12
30. Ravi M, Hopfinger AJ, Hormann RE, Dinan L (2001) 4D-QSAR analysis of a set of ecdysteroids and a comparison to CoMFA modeling. J Chem Inf Comput Sci 41:1587–1604
31. Krasowski MD, Hong X, Hopfinger AJ, Harrison NL (2002) 4D-QSAR analysis of a set of propofol analogues: mapping binding sites for an anesthetic phenol on the GABAA receptor. J Med Chem 45:3210–3221
32. Hong X, Hopfinger AJ (2003) 3D-pharmacophores of flavonoid binding at the benzodiazepine GABAA receptor site using 4D-QSAR analysis. ChemInform 34
33. Golbraikh A, Tropsha A (2002) Beware of q2! J Mol Graph Model 20:269–276
34. Tropsha A, Gramatica P, Gombar V (2003) The importance of being earnest: validation is the absolute essential for successful application and interpretation of QSPR models. QSAR Comb Sci 22:69–77
35. Tong W, Hong H, Xie Q, Shi L, Fang H, Perkins R (2005) Assessing QSAR limitations-a regulatory perspective. Curr Comput Aided Drug Des 1:195–205
36. OECD Principles for the Validation of (Q)SARs (2004). https://www.oecd.org/chemicalsafety/risk-assessment/37849783.pdf. Accessed 30 Jul 2021
37. Waterbeemd H, van de Carter RE, Grassy G, Kubinyi H, Martin YC, Tute MS, Willett P (1998) Glossary of terms used in computational drug design (IUPAC recommendations 1997). In: Annual reports in medicinal chemistry. Elsevier, pp 397–409
38. Khlebnikov AI, Schepetkin IA, Domina NG, Kirpotina LN, Quinn MT (2007) Improved quantitative structure-activity relationship models to predict antioxidant activity of flavonoids in chemical, enzymatic, and cellular systems. Bioorg Med Chem 15:1749–1770
39. Papa E, Dearden JC, Gramatica P (2007) Linear QSAR regression models for the prediction of bioconcentration factors by physicochemical properties and structural theoretical molecular descriptors. Chemosphere 67:351–358
40. Mutihac L, Mutihac R (2008) Mining in chemometrics. Anal Chim Acta 612:1–18
41. Ivanenkov YA, Savchuk NP, Ekins S, Balakin KV (2009) Computational mapping tools for drug discovery. Drug Discov Today 14:767–775
42. Zuvela P, David J, Yang X, Huang D, Wong MW (2019) Non-linear quantitative structure–activity relationships modelling, mechanistic study and in-silico design of flavonoids as potent antioxidants. Int J Mol Sci 20(9):2328
43. Zhou W, Wu S, Dai Z, Chen Y, Xiang Y, Chen J, Sun C, Zhou Q, Yuan Z (2015) Nonlinear QSAR models with high-dimensional descriptor selection and SVR improve toxicity prediction and evaluation of phenols on *Photobacterium phosphoreum*. Chemometr Intell Lab Syst 145:30–38
44. Ramandi M, Riahi S, Rahimi H, Mohammadi-Khanaposhtani M (2020) Molecular docking, linear and nonlinear QSAR studies on factor Xa inhibitors. Struct Chem 31:2023–2040
45. Breiman L, Friedman J, Olshen RA, Stone CJ (1984) Classification and regression trees Chapman & Hall
46. Breiman L (2001) Random forests. Mach Learn 45:5–32
47. He L, Jurs PC (2005) Assessing the reliability of a QSAR model's predictions. J Mol Graph Model 23:503–523
48. Roy PP, Leonard JT, Roy K (2008) Exploring the impact of size of training sets for the development of predictive QSAR models. Chemometr Intell Lab Syst 90:31–42
49. Von Der Ohe PC, Kühne R, Ebert RU, Schüürmann G (2007) Comment on "Discriminating toxicant classes by mode of action: 3. Substructure indicators" (M. Nendza and M. Müller, SAR QSAR Environ. Res. 18 155 (2007)). SAR QSAR in Environ Res 18:621–624

50. Efron B (1979) Bootstrap methods: another look at the jackknife. Ann Stat 7:1–26
51. Ravichandran (2016) Exploring the structural insights of indole-7-carboxamides as anti-HIV agents. FARMACIA 64:745–756
52. Veerasamy R, Rajak H, Jain A, Sivadasan S, Varghese CP, Agrawal RK (2011) Validation of QSAR models-strategies and importance. Int J Drug Des Discov 3:511–519
53. Chirico N, Papa E, Gramatica P (nd) Externally predictive QSAR models: thresholds of acceptance by various external validation criteria and critical inspection of scatter plots. http://www.cadaster.eu/sites/cadaster.eu/files/workshop/1_19.pdf. Accessed 30 Jul 2021
54. Hawkins DM, Basak SC, Mills D (2003) Assessing model fit by cross-validation. J Chem Inf Comput Sci 43:579–586
55. Hawkins DM (2004) The problem of overfitting. J Chem Inf Comput Sci 44:1–12
56. Norinder U (1996) Single and domain mode variable selection in 3D QSAR applications. J Chemom 10:95–105
57. Kubinyi H (2007) A general view on similarity and QSAR studies. Computer-Assisted lead finding and optimization. Verlag Helvetica Chimica Acta, Zürich, pp 7–28
58. Kubinyi H, Hamprecht FA, Miętzner T (1998) Three-dimensional quantitative similarity—activity relationships (3d qsiar) from seal similarity matrices. J Med Chem 41:2553–2564
59. Dearden JC, Cronin MTD, Kaiser KLE (2009) How not to develop a quantitative structure-activity or structure-property relationship (QSAR/QSPR). SAR QSAR Environ Res 20:241–266
60. Nantasenamat C, Isarankura-Na-Ayudhya C, Naenna T, Prachayasittikul V (2009) A practical overview of quantitative structure-activity relationship. EXCLI J 8:74–88
61. Verma RP, Hansch C (2005) An approach toward the problem of outliers in QSAR. Bioorg Med Chem 13:4597–4621
62. Kim KH (2007) Outliers in SAR and QSAR: 2. Is a flexible binding site a possible source of outliers? J Comput Aided Mol Des 21:421–435
63. Sahigara F, Ballabio D, Todeschini R, Consonni V (2013) Defining a novel k-nearest neighbours' approach to assess the applicability domain of a QSAR model for reliable predictions. J Cheminform 5:27
64. Kar S, Roy K, Leszczynski J (2018) Applicability domain: a step toward confident predictions and decidability for QSAR modeling. Methods Mol Biol 1800:141–169
65. Guha R (2021) CDK Descriptor Calculator GUI (version 1.4. 6). http://www.rguha.net/code/java/ cdkdesc.html. Accessed 30 Jul 2021
66. rcdk: Interface to the CDK Libraries. https://cran.r-project.org/web/packages/rcdk/index.html. Accessed 30 Jul 2021
67. Yap C (2021) PaDEL-Descriptor. http://www.yapcwsoft.com/dd/padeldescriptor. Accessed 30 Jul 2021
68. Yap CW (2011) PaDEL-descriptor: an open-source software to calculate molecular descriptors and fingerprints. J Comput Chem 32:1466–1474
69. Xiao N, Cao D, Xu QR (2021) Toolkit for compound-protein interaction in drug discovery. http://bioconductor.org/packages/release/bioc/html/Rcpi.html. Accessed 30 Jul 2021
70. Moplex Ltd (2021) Sykora V Chemical Descriptors Library. https://sourceforge.net/projects/cdelib/. Accessed 30 Jul 2021
71. Cao D (2017) ChemoPy descriptor calculator. http://www.scbdd.com/chemopydesc/index/. Accessed 30 Jul 2021
72. O'Boyle N, Banck M, James C, Morley C, Vandermeersch T, Hutchison G (2017) Open Babel: the open source chemistry toolbox. http://openbabel.org/. Accessed 30 Jul 2021
73. Moriwaki H, Tian Y, Kawashita N (2018) Mordred: a molecular descriptor calculator. J Cheminform 10(4). https://doi.org/10.1186/s13321-018-0258-y
74. RDKit (2021) Open-source cheminformatics. http://www.rdkit.org. Accessed 30 Jul 2021
75. Nantasenamat C, Worachartcheewan A, Jamsak S, Preeyanon L, Shoombuatong W, Simeon S, Mandi P, Isarankura-Na-Ayudhya C, Prachayasittikul V (2015) AutoWeka: toward an automated data mining software for QSAR and QSPR studies. Methods Mol Biol 1260:119–147

76. Staalring JC, Carlsson LA, Almeida P, Boyer S (2011) AZOrange-High performance open-source machine learning for QSAR modeling in a graphical programming environment. J Cheminform 3:1–10

77. Kuhn T, Willighagen EL, Zielesny A, Steinbeck C (2010) CDK-Taverna: an open workflow environment for cheminformatics. BMC Bioinform 11:159

78. Miller BT, Singh RP, Klauda JB, Hodoscek M, Brooks BR, Woodcock HL 3rd (2008) CHARMMing: a new, flexible web portal for CHARMM. J Chem Inf Model 48:1920–1929

79. Walker T, Grulke CM, Pozefsky D, Tropsha A (2010) Chembench: a cheminformatics workbench. Bioinformatics 26:3000–3001

80. Backman TWH, Cao Y, Girke T (2011) ChemMine tools: an online service for analyzing and clustering small molecules. Nucl Acids Res 39:W486–W491

81. Benfenati E, Toropov AA, Toropova AP, Manganaro A, Gonella Diaza R (2011) Coral software: QSAR for anticancer agents: Coral Software: QSAR for anticancer agents. Chem Biol Drug Des 77:471–476

82. DTAI Research Group (2017) DMax chemistry assistant. https://dtai.cs.kuleuven.be/software/dmax/. Accessed 30 Jul 2021

83. Sushko I, Novotarskyi S, Körner R et al (2011) Online chemical modeling environment (OCHEM): web platform for data storage, model development and publishing of chemical information. J Comput Aided Mol Des 25:533–554

84. Dimitrov SD, Diderich R, Sobanski T, Pavlov TS, Chankov GV, Chapkanov AS, Mekenyan OG (2016) QSAR Toolbox–workflow and major functionalities. SAR QSAR Environ Res 27:203–219

85. Filimonov DA, Lagunin AA, Gloriozova TA, Rudik AV, Druzhilovskii DS, Pogodin PV, Poroikov VV (2014) Prediction of the biological activity spectra of organic compounds using the pass online web resource. Chem Heterocycl Compd (NY) 50:444–457

86. Gramatica P, Chirico N, Papa E, Cassani S, Kovarich S (2013) QSARINS: A new software for the development, analysis, and validation of QSAR MLR models. J Comput Chem 34:2121–2132

87. Cox R, Green DVS, Luscombe CN, Malcolm N, Pickett SD (2013) QSAR workbench: automating QSAR modeling to drive compound design. J Comput Aided Mol Des 27:321–336

88. Patlewicz G, Jeliazkova N, Safford RJ, Worth AP, Aleksiev B (2008) An evaluation of the implementation of the Cramer classification scheme in the Toxtree software. SAR QSAR Environ Res 19:495–524

89. Martins JPA, Barbosa EG, Pasqualoto KFM, Ferreira MMC (2009) LQTA-QSAR: a new 4D-QSAR methodology. J Chem Inf Model 49:1428–1436

90. Halder AK, Dias Soeiro Cordeiro MN (2021) QSAR-Co-X: an open-source toolkit for multitarget QSAR modelling. J Cheminform 13(29). https://doi.org/10.1186/s13321-021-00508-0

91. Drug Theoretics and Cheminformatics (DTC) laboratory software tools. https://dtclab.webs.com/software-tools. Accessed 30 Jul 2021

92. Soufan O, Ba-alawi W, Magana-Mora A (2021) DPubChem: a web tool for QSAR modeling and high-throughput virtual screening. Sci Rep. https://www.nature.com/articles/s41598-018-27495-x. Accessed 30 Jul 2021

93. Enciso M, Meftahi N, Walker ML, Smith BJ (2016) BioPPSy: an open-source platform for QSAR/QSPR analysis. PLoS ONE 11:e0166298

Simulating Landscape Hydrologic Connectivity in a Precise Manner Using Hydro-Conditioning

Rallapalli Srinivas, Matt Drewitz, Joe Magner, Ajit Pratap Singh, Dhruv Kumar, and Yashwant Bhaskar Katpatal

Abstract High resolution Light Detection and Ranging (LiDAR)-derived Digital Elevation Models (DEMs) have significantly enhanced hydrological modeling and agricultural planning. However, it is important to accurately simulate the landscape flow network by modifying the LiDAR derived DEM. Hydro-conditioning identifies false pools and depressions and places breach lines to ensure continuous flow through the surfaces such as bridges, culverts, and railroads. This study explores the variations in the criteria namely flow network, impeded flows, depression etc. which determine the finest locations of best management practices using watershed models. Study compares different levels of hydro-conditioned DEMs for the Plum Creek sub-watershed, Minnesota. Results indicate that both manual and automated 'hDEMs' facilitate field scale planning and practice siting to different degrees. Outcomes help in planning cost-effective precision agriculture activities.

Keywords Agricultural planning · Hydro-conditioning · Hydrology · LiDAR

R. Srinivas (✉) · A. P. Singh
Department of Civil Engineering, Birla Institute of Technology and Science, Pilani, India
e-mail: r.srinivas@pilani.bits-pilani.ac.in

M. Drewitz
Board of Water and Soil Resources, Saint Paul, MN, USA

J. Magner
Department of Bioproducts and Biosystems Engineering, University of Minnesota, Saint Paul, MN, USA

D. Kumar
Department of Computer Science and Engineering, University of Minnesota, Minneapolis, MN, USA

Y. B. Katpatal
Department of Civil Engineering, Visvesvaraya National Institute of Technology, Nagpur, India

© The Author(s), under exclusive license to Springer Nature Singapore Pte Ltd. 2022 209
R. Srinivas et al. (eds.), *Advances in Computational Modeling and Simulation*,
Lecture Notes in Mechanical Engineering,
https://doi.org/10.1007/978-981-16-7857-8_17

1 Introduction

Simulation and modeling of hydrology across the landscape is pivotal for any watershed modeling [1]. The transport and fate of the pollutants in the agricultural watersheds is very much dependent on the proper hydrologic simulation [2, 3]. Success of robust watershed models is a function of usage of high-resolution LiDAR based Digital Elevation Models [4, 5]. Recently, hydrologic modelers observed that despite the availability of high-resolution LiDAR images of digital elevation models, it is a challenge for LiDAR sensors to capture the hydrologic processes happening under the earth's surfaces like (bridges, railroads, culverts, etc.). As a result, flow direction/accumulation network generated using such a DEM is not simulating the flow with accuracy [6, 7].

Modelers and planners are spending technical and economic resources toward a new technology namely hydro-conditioning (hydro-enforcement) which is capable of modifying the existing LiDAR derived DEM to accurately map the flow of water under or through the various features on the earth surfaces [8]. Without hydro-enforcement, natural flow of water is obstructed in the DEM and roads would appear as dams and there will be false pooling on the upstream side. Such faulty representation has a detrimental effect on the outputs of the watershed models which use these un-hydro conditioned DEMs [9]. These days, watershed models play a critical role in estimating the optimal locations of BMPs in the agricultural field. To obtain accurate outcomes at a field-scale, hydro-conditioning is required as it performs a thorough terrain analysis, accurate calculations of runoff parameters. Ultimately, stream flow network is simulated to replicate landscape hydrology with least errors.

There are different levels of hydro-conditioning depending on the available time, capital and expertise. It also depends on the goals of a watershed planning and management committee. Broadly these are termed as manual and automated hydro-conditioning. In this study, we try to assess the impact of using manually and automatically generated hydro-conditioned DEMs on the criteria such as stream flow network, impeded/false flows, boundaries of the catchments, and true/false depressions. All these criteria are used by a conservation precision technologies or watershed models such as Agricultura Conservation Planning Framework (ACPF), Prioritize Target and Measure Application [1, 10]. A case of Plum creek watershed in Minnesota, USA, has been demonstrated.

2 Manual and Automated Hydro-Conditioning Process

Hydro-conditioning demands extensive knowledge of watershed characteristics as well as technical and modeling skills. Moreover, it is an expensive process as we move toward better accuracy. The level of accuracy and detail in the hydro-conditioning depends on the purpose and specific needs of the watershed committee and decision makers [7]. In this study, we will try to understand that how manually (HEI, 2016) and automatically Gelder [6] generated hydro-conditioned DEMs affect the flow network of a watershed. The detailed process of generating these DEMs is given in literature [6, 8]. Tables 1 and 2 highlight the major utilities and difference of these DEMs.

Table 1 Utilities manual and automated hydro-conditioned DEMs

Watershed area	Flow and load routing for field scale BMP siting, preliminary BMP design	Flow and load routing at field scale BMP siting	Flow and load routing at watershed scale + Lake analysis	Flow and load routing at watershed scale	Delineating watershed+ Basic terrain analysis
Single field (≤ 40 acres)					
Public Land Survey System (40 acres to 1 sq. miles)	Planning and implementation at field scale-Generated using manual process (HE, 2016)				
Sub-watershed (1 to 50 sq. miles)			Planning, especially for Daily Erosion Project - Generated using automated process (Gelder, 2015)		
Major watershed (> 50 sq. miles)					

Table 2 Description of manual and automated hydro-conditioned DEMs

Type of hydro-conditioning	Description
Generated using automated process (Gelder 2015)	Pit-filling and Hole punching done Impediments resulting from intersections of perennial watercourse crossings and major river conveyance landforms are removed from the DEM Impediments associated with DNR mapped watercourses are removed from the hDEM in a watershed of interest Automated code is written to identify true depressions
Generated using manual process (HEI 2016)	In addition to automated corrections, All impediments affecting modeled flow and proper flow path develop are removed from the hDEM throughout a watershed of interest. In addition, it is ensured that 'true depressions' are maintained in the DEM Manual verification is done

3 Results

Purpose of both manual and automated hDEM is to identify and accurately represent the impeded flows, true depressions, and flow network. These parameters are necessary for delineation of BMP sites. Adequate simulation and modeling of hydrology (flow routing) is a function of algorithm used for identification of impeded flows and true/false depressions. Impeded flows are usually observed at the obstructions (such as roads), which create inaccurate flow paths (ponding effect and spilling over-flow). Using the manual and automated DEMs, impeded flows are generated in ArcGIS module (Fig. 1) for Plum creek watershed in Minnesota. Figure 1 shows that manual 'hDEM' has comparatively lesser number of impeded flows as compared to automated one. This is because there are optimum number of field-verified cut lines placed in manual hydro-enforcement. Moreover, automated hDEM also ensures that true depressions on the landscape are maintained in order to replicate the landscape hydrology as accurately as possible. Figure 2 shows the variation in the number of false depressions, which reduce with high-level enforcement and finally, 'manual hDEM' represents only true depressions. The number of depressions reduce to 160 (manual hDEM) from 655 (automated hDEM), which shows a significant reduction in false depressions (Table 3). It is also interesting to note that the maximum depth of depressions in manual hDEM is 178 cm as compared to 285 cm for automated hDEM. The total sum of depth of depressions also varied from 35,509 to 12,093 cm.

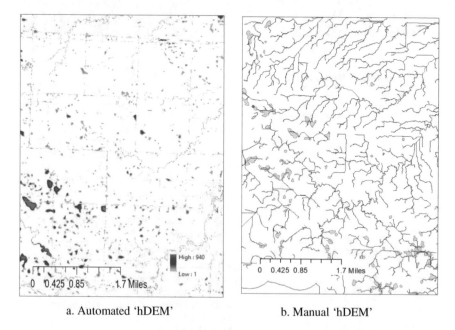

a. Automated 'hDEM'   b. Manual 'hDEM'

Fig. 1 Representation of impeded flows for automated and manual hydro-conditioning

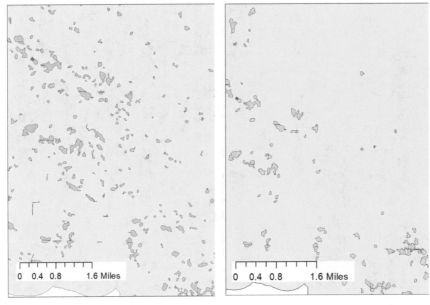

a. Automated 'hDEM'   b. Manual 'hDEM'

Fig. 2 Depressions in automated and manual hydro-conditioning

Table 3 Depressions corresponding to hydro-conditioning

Attribute	Automated	Manual
Count	655	160
Min depth (cm)	30	30
Max depth (cm)	285	178
Sum (cm)	35,509	12,093

These numbers give a clear indication to the watershed modelers that how hydro-conditioning can have a significant impact on impeded flows and depressions and thus, the flow network simulation.

Simulated hydrology (flow network) generated using automated and manual hDEMs are represented in Fig. 3. Flow detailing as well as continuity significantly varies with hydro-conditioning levels. The improved stream network in manual hDEM is due to proper breaching of the impeded flows with breach lines to ensure hydrologic connectivity, which is essential for generating catchments as well as for siting BMPs especially riparian practices. Manual verification improves the flow network interpretation. The total count of flow lines varies from 1741 (automated) to 2090 (manual) due to placement of an optimal number of breach (cut/burn) lines resulting in a total sum of 823,923.2 m of flow lines in a 'manual hDEM' (Table 4). Watershed models focus on siting BMPs such as saturated buffers and riparian

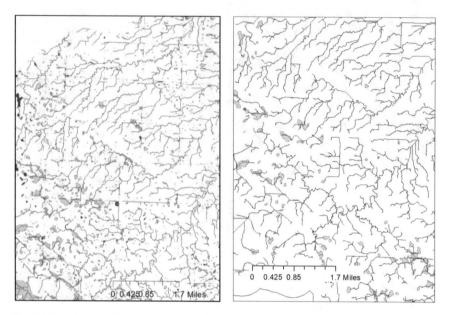

Fig. 3 Flow network of automated (left) and manual (right) hydro-conditioned DEM

Table 4 Statistics related to flow network based on hydro-conditioning

Attribute	Automated	Manual
Count	1,741	2,090
Count/km2	6	7
Min length (m)	3	2.12
Max length (m)	2,495.42	2,730.57
Sum (m)	711,927	823,923.2

functions which render ecological benefits along with enhanced denitrification. Thus, accurate hydrologic representation is a key.

4 Conclusions

The study compares manual and automated hDEMs in terms of their ability in replicating the landscape hydrology which is crucial for application of watershed models. The results indicate that both hDEMs delineate hydrology with different accuracy; however, manual hDEMs have an advantage as the DEM is continuously enhanced based on manual field verification. On the other hand, automated hDEM has an automatic code which can differentiate between true and false depressions with good accuracy. The study therefore proposes the integration of both technologies to have better interpretation and simulation of landscape hydrology.

References

1. Srinivas R, Drewitz M, Magner J (2020) Evaluating watershed-based optimized decision support framework for conservation practice placement in Plum Creek Minnesota. J Hydrol 583:124573
2. Poppenga SK, Gesch DB, Worstell BB (2013) Hydrography change detection—the usefulness of surface channels derived from LiDAR DEMS for updating mapped hydrography: J Am Water Res Assoc 49(2), 371–389
3. Martins VS, Kaleita AL, Gelder BK, da Silveira HL, Abe CA (2020) Exploring multi-scale object-based convolutional neural network (multi-OCNN) for remote sensing image classification at high spatial resolution. ISPRS J Photogramm Remote Sens 168:56–73
4. Epelde AM, Antiguedad I, Brito D, Jauch E, Neves R, Garneau C, Sánchez-Pérez JM (2016) Different modelling approaches to evaluate nitrogen transport and turnover at the watershed scale. J Hydrol 539:478–494
5. Srinivas R, Singh AP, Dhadse K, Garg C, Deshmukh A (2018) Sustainable management of a river basin by integrating an improved fuzzy based hybridized SWOT model and geo-statistical weighted thematic overlay analysis. J Hydrol 563:92–105
6. Gelder BK (2015) Automation of DEM cutting for hydrologic/hydraulic modeling. In: Trans Proj Rep 103
7. Vaughn SR (2017) DEM Hydro-modification: a composition to help understand its necessity. Technical manuscript. Minnesota information services at Minnesota Department of Natural Resources—Ecological and Water Resources
8. HEI (2016) Hydrologic conditioning and terrain analysis report. Houston Engineering Inc. (HEI), Maple Grove, MN, 8249, 001
9. Cochrane TA, Yoder DC, Flanagan DC, Dabney SM (2019) Quantifying and modeling sediment yields from interrill erosion under armouring. Soil Tillage Res 195:104375
10. Porter SA, Tomer MD, James DE, Boomer KMB (2018) Agricultural conservation planning framework ArcGIS toolbox user's manual. National laboratory for agriculture & the environment, USDA-ARS Ames, Iowa

The Effect of Matrix Property Smoothing on the Reliability of Fibre-Reinforced Composites

Sadik L. Omairey, Peter D. Dunning, and Srinivas Sriramula

Abstract This paper presents the development and assessment of a smoothing technique for the random properties of the matrix phase of a fibre-reinforced composite as part of a multiscale reliability framework. Many sectors in the industry are using fibre-reinforced composite materials, utilising their high stiffness-to-mass density ratio. Still, many uncertainties occur in their properties because of their multiscale build-up nature. Hence, structural reliability analysis can produce efficient designs, but it requires an understanding of how all sources of uncertainty affect performance. Among other approaches, the authors developed a multiscale surrogate-based framework for reliability analysis, which uses large representative volume elements that can correlate and propagate the effect of several uncertainties, including a spatial variation of matrix properties. The framework uses a blur kernel to smooth matrix properties. In this study, the effect of using the blur kernel filter and other larger filters is examined in terms of their impact on the statistical properties of the matrix phase and the overall stiffness reliability of an analytical laminate example. The developed kernels proportionally reduce the standard deviation of the matrix property towards a lower limit as they increase in size. It is also shown that the stiffness reliability of the selected composite example (which includes the chosen parameters for laminate configuration, loading type, material, and other uncertainties in the system) is not affected by the use of the matrix smoothing technique.

Keywords Fibre-reinforced composites · Matrix · Homogenisation · Uncertainty · Reliability

S. L. Omairey (✉)
Brunel Composites Centre, College of Engineering, Design and Physical Sciences, Brunel University London, London, UK
e-mail: sadik.omairey@brunel.ac.uk

P. D. Dunning · S. Sriramula
School of Engineering, University of Aberdeen, Aberdeen, UK

© The Author(s), under exclusive license to Springer Nature Singapore Pte Ltd. 2022
R. Srinivas et al. (eds.), *Advances in Computational Modeling and Simulation*,
Lecture Notes in Mechanical Engineering,
https://doi.org/10.1007/978-981-16-7857-8_18

1 Introduction

Utilisation of composite materials in various industries is on the rise, replacing alloys in many areas, benefiting from their higher stiffness-to-weight ratio. From a technical perspective, the multi-material build-up of fibre-reinforced materials permits numerous uncertainties to occur across the material scales, which often is not the case in homogeneous alloys [1–8]. Therefore, to design composites effectively, there is a need to consider the effect of composite uncertainties and propagate their influence between the material scales to clearly understand the overall composite properties. This modelling strategy could lead to reliable designs and optimum use of the material [9–12]. Monte Carlo simulation (MCS) can be used to achieve this probabilistic design approach; in which MCS generates sets of random uncertainties based on their corresponding statistical data. The generated random variables are then propagated through the material scales using homogenisation models to assess their influence on the composite properties at the laminate or component scale. Homogenisation methods are broadly divided into two categories, analytical and numerical. The last is commonly used as it permits examining many design variables compared with analytical methods that are often hindered by assumptions and limitations [12–14].

FEA-based homogenisation strategy requires defining a Representative Volume Element (RVE). Using detailed and large-scale RVEs allow representing many uncertainties, including spatial variations. However, it can increase the computational cost and limit the possibility of constructing a probabilistic analysis framework. Therefore, to have a numerical framework capable of analysing many RVEs as part of the MCS reliability analysis, efficient surrogate models are employed to replace FEA simulation with analytical approximations [14–16]. For example, previous work by the authors [12, 17, 18] established a probabilistic framework that use efficient FEA-based surrogate models to calculate the stiffness properties of UD composite laminates considering uncertainties at micro, meso, and laminate scales. This framework can capture more uncertainties between the micro and laminate scale compared with other methods as it developed an inclusive large representative volume element (LRVE). The LRVE represents a larger section of uncertainties associating random RVEs through several spatial correlation principles. The LRVE accounts for the micro- and meso-scale uncertainties, including fibre stacking, uncertain fibre cross-sectional areas, and material properties of the fibre-matrix constituents. For the latter, matrix phase within the composite is an uninterrupted media, unlike fibres. Consequently, having sharp changes in its property values is unrealistic; Any change in its property value should be gradual. However, few studies evaluating matrix property variation at this scale are available to support this theory; One of which is a study by Riaño et al. [19]; this study used atomic force microscopy (AFM) to determine the thickness and the elastic modulus of the matrix–fibre interphase region, assuming the rest of the matrix to be uniform. In the first 0.2 μm from the fibre/matrix interface, the modulus showed an approximate decrease of 35% compared with the average matrix modulus obtained within the scanned 1.0 μm frame.

In authors' previous work, an image processing concept was used to smooth out the randomness of matrix material properties within the LRVE. In this study, the influence of the property smoothing approach used in the LRVE is investigated using different blur kernels and observing their effect on the statistical properties of the matrix phase and the overall stiffness reliability of an analytical composite laminate example. Although the developed kernels are not backed by experimental data, frame size and kernel weights allow customisation if such data became available.

In Sect. 2 of this paper, the methodology of the study is explained. Then, Sect. 3 demonstrates and discusses applying the developed kernels to an analytical composite laminate example. Finally, Sect. 4 highlights the key findings of the study.

2 Methodology

Assessing composite component reliability at a design point requires propagating the influence of uncertainties across the component scales. To achieve this, the authors have previously developed an efficient FEA-based multiscale probabilistic framework that generates an offline library of probabilistic material properties [17]. This probabilistic framework comprises of FEA-based, computationally efficient surrogate models, capable of capturing the effect of various multiscale uncertainties on the homogenised elastic properties of continuous fibre-reinforced composites. It takes the statistical information of uncertainties as an input and uses the LRVE that can model spatial variation of uncertainties because of its size, compared with a single RVE which contains only one or two fibres, as shown in Fig. 1. Computational efficiency using the LRVE is achieved by employing a string of surrogate models trained using a reduced number of FE data points. The data points are obtained using the periodic RVE homogenisation method [20]. This approach makes it feasible to assess the

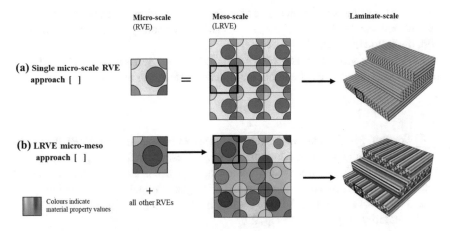

Fig. 1 The structure of homogenisation frameworks adopted to examine matrix property smoothing

reliability accurately using MCS with a large amount of probabilistic homogenised properties.

The spatial correlation of the RVEs that form the LRVE is considered in the developed framework to precisely describe a larger portion of the material at the meso scale by assembling defined number of micro-scale RVEs. Namely, fibres of neighbouring RVEs are correlated with respect to their properties and geometric arrangements. As for the matrix, the LRVE uses an image processing concept to smooth out the matrix property randomness and avoid unrealistic jumps between neighbouring RVEs. In our previous work, a 3 × 3 smoothing blur kernel was used to smooth the random matrix stiffness properties of each RVE within the LRVE. In this study, the effect of the blur kernel on computed reliability is investigated. The configuration of the representative units' and the different kernels developed in this study are illustrated in the following sections.

2.1 Single Micro-Scale RVE

In the previous work [12], the reliability of composite laminates was computed using homogenised stiffness properties of randomly generated RVEs, where no spatial variation was considered for the uncertain parameters within a single lamina ply. Thus, the properties of each lamina are derived from a single 1 × 1 random RVE, see Fig. 1a. This model is used in this study as a baseline to compare against the LRVE model with different matrix blur kernels.

2.2 LRVE and Blur Kernels

Adopting LRVEs allows an inclusive material representation, describing more uncertainties between the micro and laminate scale. In this study, 8 × 8 LRVEs are used with the following kernels:

$^{1\cdot1}$**kernel (without smoothing)**: The randomly generated matrix stiffness properties are directly used without any filtering effect, see Eq. 1 and Fig. 2a.

$$^{1\cdot1}kernel = [1] \tag{1}$$

$^{3\cdot3}$**kernel**: Matrix material properties for an RVE (within the LRVE) are calculated as the average of the random matrix properties within a 3 × 3 grid centred on the RVE, see Eq. 2. It is important to note that there is a need to generate additional matrix random values surrounding the LRVE edges to populate the grid at all positions; hence, for an 8 × 8 LRVE, we start with a 10 × 10 grid of randomly generated matrix material properties, see Fig. 2b.

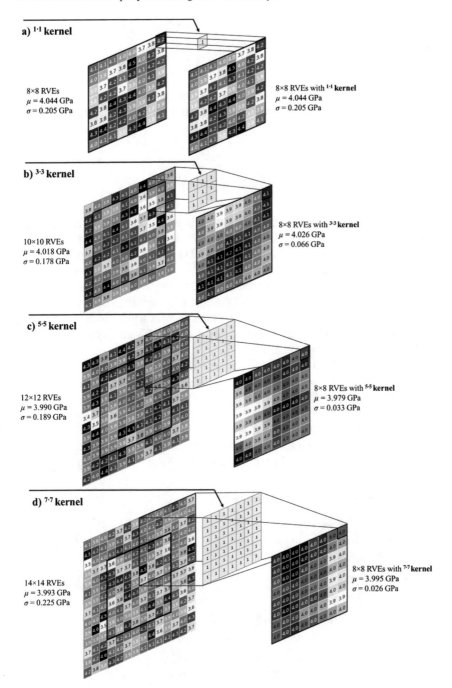

Fig. 2 The different sizes of blur kernels used in this study to smooth random matrix stiffness of 8 × 8 LRVEs

$$^{3 \cdot 3}kernel = \frac{1}{9}\begin{bmatrix} 1 & 1 & 1 \\ 1 & 1 & 1 \\ 1 & 1 & 1 \end{bmatrix} \tag{2}$$

$^{5 \cdot 5}$**kernel and**$^{7 \cdot 7}$**kernel**: These kernels work the same as the $^{3-3}$kernel, but average the matrix material properties over a larger area: 5×5 and 7×7 RVE grids, respectively, see Fig. 2c, d.

3   Example

The effect of using different matrix blur kernels is assessed using an analytical laminate example. The laminate is made of symmetrically arranged set of four specially orthotropic 0.5 mm thick laminas (see Fig. 3). For which, buckling stiffness reliability is selected for the assessment. The properties of each lamina are obtained from the random RVEs and LRVEs. Thus, the classical lamination theory is used estimates the laminate properties using four lamina stiffness properties in each MCS sample. On top of material properties and geometry uncertainties within the micro and meso scale, other uncertainties such as lamina ply thickness (t) and ply orientation (θ_p) are considered as illustrated in Sect. 3.1.2.

The limit state function (LSF) calculations of the specially orthotropic laminate configuration adopted are simplified as this laminate example only requires stiffness components D_{11}, D_{12}, D_{22}, and D_{66}. Uncertainties are randomly generated using MCS; these are then used to compute stiffness properties using the homogenisation surrogate models. D_{ij} stiffness terms along with the effect of laminate-scale uncertainties are calculated with classical lamination theory [21, 22]. Finally, the LSF for each randomly generated laminate is evaluated to assess failure probability as

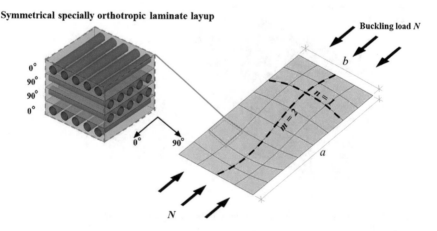

Fig. 3 The selected buckling loading condition for a symmetrically specially orthotropic laminate

serviceability limits are attained: $P_f = P[g(X) \leq 0]$, see Eq. 3 [23].

$$g_b(X) = N - N_{LS} = \pi^2 \left[\begin{matrix} D_{11}\left[\dfrac{m}{a}\right]^2 + 2(D_{12} + 2D_{66})\left[\dfrac{n}{b}\right]^2 \\ +D_{22}\left[\dfrac{n}{b}\right]^4\left[\dfrac{a}{m}\right]^2 \end{matrix} \right] - N_{LS} \qquad (3)$$

The above applies for a simply supported symmetrical specially orthotropic laminate, where:

N_{LS}:	Buckling load along the width of the laminate (155N/mm)	m, n:	The number of half wavelengths at $0°$ and $90°$-direction, respectively, $m = 2$ and $n = 1$ for buckling
D_{ij}:	Laminate stiffness components	a, b:	Laminate length (100mm), and width (50mm), respectively

3.1 Propagated Uncertainties

In this study, E-glass fibre-epoxy composite is selected as the material of the laminate. Uncertainties for this composite material system are introduced within two categories: the micro and meso scales, and laminate scale. These two categories are briefly explained below, and their detailed implementation can be found in the authors' previous work [17].

3.1.1 Micro and Meso Scales Uncertainties

Material uncertainties: These uncertainties are represented by varying the stiffness properties of each phase based on their statistical properties. For a single RVE, this includes one matrix and five fibre sections (central fibre and four quarters), as shown in Fig. 1a.

Fibre stacking and volume ratio uncertainties: The developed reliability framework splits fibres into fixed and non-fixed fibres. RVEs' four fibre corner quarters represent the fixed fibres; in which these quarters remain in place and have the same diameter to preserve connection with its neighbouring RVE. On the other hand, the central non-fixed fibres can shift within the RVE to model stacking (r and θ) and V_f ratio uncertainties without violating RVE's boundary periodicity.

In the case of the single RVE, designating material property uncertainties is straightforward as each sample in the MCS uses a separate RVE. But in the case of the LRVE, two types of correlations are used:

Table 1 The material properties and their statistical information for the E-glass fibre-epoxy composite

Property		Mean/lower limit	Distribution	CoV/limits	Categories
Fibre (E-glass)	E_m(GPa)	72.45	Normal	5%	Micro and Meso scales
	v_m(ratio)	0.25	Normal	5%	
	Fibre-volume ratio V_f	0.52	Normal	5%	
	Fibre stacking (r and θ)	RVE centre, $0°$	Uniform	r: 0–8%[a], θ: $0°$-$360°$	
Matrix (Epoxy)	E_f(GPa)	4.0	Normal	5%	
	v_f(ratio)	0.3	Normal	5%	
Lamina	ply thickness (t)	0.5 mm	Normal	5%	Laminate scale
	ply orientation (θ_p)	$[0°/90°]_S$	Normal	5 °	

[a] *Percentage of the RVE edge length*

Fibre phase properties: Neighbouring RVEs within each LRVE are correlated by assigning the same fibre material property values to all adjacent fixed corner fibre, forming a link between individual RVEs and their surroundings.

Matrix phase property uncertainties: The proposed blur kernels listed in Sect. 2.2 are used to avoid having unrealistic sharp changes in matrix material properties, as it is a continuous media.

3.1.2 Laminate Scale Uncertainties

This category of uncertainties is within the laminate scale; it includes uncertainty in both ply thickness and ply orientation.

The properties and their associated statistical information for the example used in this study are listed in Table 1, where the distributions are similar to those used in previous studies [24–27].

3.2 Impact of Kernels on Statistical Distribution of the Matrix Properties

Before examining the effect of using different matrix blur kernels on reliability, their impact on the statistics of the matrix properties within the LRVEs is investigated. This is done by plotting the distribution of smoothed matrix stiffness values for 10,000 LRVEs (which sums to 640,000 RVEs, or data points for each kernel), as seen in the shaded regions within Fig. 4. This data is also fitted with normal distribution

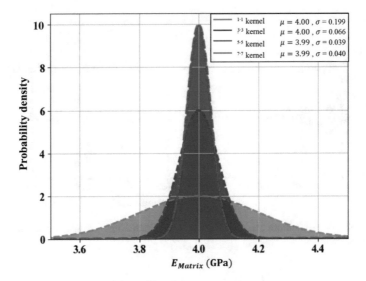

Fig. 4 Distribution of 640,000 normalised matrix property data points and fitted normal distribution curves

curves (the dotted lines in Fig. 4), showing that the data of all kernels follow normal distributions. Yet, the standard deviation decreases as the size of the kernel increases, which is expected as a larger number of samples are averaged within each grid, reducing the variance around the mean matrix stiffness value of 4.0 GPa. However, the data also shows that results for the $^{5-5}$kernel and $^{7-7}$kernal are almost the same, with a coefficient of variation of approximately 1%. Thus, variance reduction as the kernel size increases does not continue when the kernel is sufficiently large. This suggests that, for an 8 × 8 LRVE, results will not be significantly affected if the kernel is increased beyond a 5 × 5 grid. This is investigated further below.

3.3 Impact of Kernels on Reliability

The buckling reliability of the specially orthotropic laminate example with uncertainties illustrated in Table 1 is assessed employing the developed mutliscale framework [17] for five different configurations of material representative units: Single RVE, 8 × 8 LRVE with $^{1 \cdot 1}$kernel, $^{3 \cdot 3}$kernel, $^{5 \cdot 5}$kernel, and $^{7 \cdot 7}$kernel. The amount of MCS samples required (S) for each of the five configurations is calculated applying the relation shown in Eq. 4 [28], aiming for an initial probability of failure of $P'_f = 10^{-3}$, with a confidence level (α) of 80%, and maximum allowable error (δ_{P_f}) of 10%.

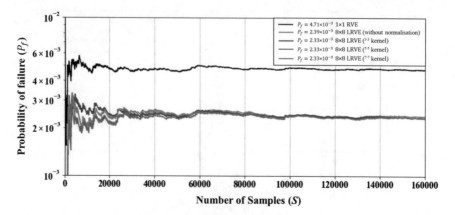

Fig. 5 Probability of failure assessment for the specially orthotropic laminate when using different sized blur kernels

$$S = \frac{P'_f\left(1 - P'_f\right)}{\delta^2_{P_f}} Z^2_{\frac{1+\alpha}{2}} = \frac{10^{-3}\left(1 - 10^{-3}\right)}{\left(1 * 10^{-4}\right)^2}(1.28)^2$$

$$\approx 160,000 \tag{4}$$

Because failure probability is a binomial distribution and $nP_f > 5$, it is acceptable to use Z to estimate the inverse of the standard normal cumulative distribution for the value $(1 + \alpha)/2$.

The first observation of the reliability results in Fig. 5 is the fact that the single RVE has a higher P_f compared with the 8×8 LRVEs. It is understood that the decrease in the P_f is because homogenised properties at the lamina-scale of the larger representative elements have less variation, consequently reducing the odds of allocating exceptionally low stiffness values to the lamina and trigger failure. As for the use of different matrix normalisation kernels within the selected 8×8 LRVE size, P_f results indicate that there is no clear impact on the reliability in this example, with all four models computing a P_f of approximately 2.3×10^{-3}. This can be explained as matrix stiffness property uncertainty is just one of many uncertainties in the system (see Table 1), and the fact that the blur kernel only affects the matrix stiffness in the LRVE, which is propagated to the laminate scale, along with the effects of all other uncertainties. Therefore, even though using a larger blur kernel reduces variance in the matrix stiffness properties (as seen in Fig. 4), it does not significantly affect the laminate scale when computing reliability. This suggests that, at least for some problems, the computed reliability is not sensitive to the choice of blur kernel size, or whether a blur kernel is even used.

4 Conclusions and Future Work

An efficient FEA-based multiscale reliability framework for continuous fibre-reinforced composites was previously developed by the authors. The framework uses a blur kernel to smooth matrix properties in the meso-scale LRVE to avoid unrealistic sharp changes as random property values are generated. Thus, the neighbouring RVEs become spatially correlated, offering a more realistic representation because the matrix phase is a continuous media. In previous work, the framework used a blur kernel with a 3×3 grid to demonstrate the possibility of implementing matrix property correlation. In this study, the influence of matrix property smoothing is further investigated using different blur kernel sizes in addition to the developed $3 \cdot 3$ kernel; these are $5 \cdot 5$ kernel and $7 \cdot 7$ kernel. Furthermore, a $1 \cdot 1$ kernel (LRVE without matrix property smoothing) and a single RVE representation are used to establish a comparison baseline.

The results show that although using larger kernels decreases the standard deviation of the smoothed matrix properties, the probability of failure remains largely unaffected by kernel choice, although there is a clear difference between using a single RVE and an 8×8 LRVE. Hence, neglecting matrix smoothing or using a $3 \cdot 3$ kernel remains valid. Nevertheless, it is essential to note that this conclusion reflects the impact seen for the selected analytical example and its parameters. Adopting another example or varying statistical information of the uncertainties can influence the sensitivity of the blur kernels. Hence, future work can investigate the effect and sensitivity of filter sizes on the reliability of complex numerical examples and correlate the use of blur frame size with either experimental-based matrix property data; or output from numerical simulations that consider the effect of fibre thermal properties on the cure (or crystallisation in the case of thermoplastic composites) of surrounding resin phase, and hence, effect on matrix mechanical properties.

References

1. González C, Lorca J (2007) Mechanical behavior of unidirectional fiber-reinforced polymers under transverse compression: microscopic mechanisms and modeling. Compos Sci Technol **67**(13):2795–2806
2. Sriramula S, Chryssanthopoulos MK (2009) Quantification of uncertainty modelling in stochastic analysis of FRP composites. Compos A Appl Sci Manuf 40(11):1673–1684
3. Komeili M, Milani AS (2012) The effect of meso-level uncertainties on the mechanical response of woven fabric composites under axial loading. Comput Struct 90–91:163–171
4. Nikopour H, Selvadurai APS (2014) Concentrated loading of a fibre-reinforced composite plate: experimental and computational modeling of boundary fixity. Compos B Eng 60:297–305
5. Yancey RN (2016) Challenges. In: In Njuguna J (ed) opportunities, and perspectives on lightweight composite structures: Aerospace versus automotive. Lightweight Composite Structures in TransportWoodhead Publishing, pp 35–52
6. Qiu S, Fuentes CA, Zhang D, Van Vuure AW, Seveno D (2016) Wettability of a single carbon fiber. Langmuir 32(38):9697–9705

7. Zhang S, Zhang L, Wang Y, Tao J, Chen X (2016) Effect of ply level thickness uncertainty on reliability of laminated composite panels. J Reinf Plast Compos 35(19):1387–1400
8. Tomar SS, Zafar S, Talha M, Gao W, Hui D (2018) State of the art of composite structures in non-deterministic framework: a review. Thin-Walled Struct 132:700–716
9. Di Sciuva M, Lomario D (2003) A comparison between monte carlo and FORMs in calculating the reliability of a composite structure. Compos Struct 59(1):155–162
10. Boyer C, Béakou A, Lemaire M (1997) Design of a composite structure to achieve a specified reliability level. Reliab Eng Syst Saf 56(3):273–283
11. Maha A, Bennett RM, Abdul-Hamid Z (2004) Probabilistic based design of concentrically loaded fiber-reinforced polymeric compression members. J Struct Eng 130(12):1914–1920
12. Omairey SL, Dunning PD, Sriramula S (2018) Influence of micro-scale uncertainties on the reliability of fibre-matrix composites. Compos Struct 203:204–216
13. Cheng G, Cai Y, Xu L (2013) Novel implementation of homogenization method to predict effective properties of periodic materials. Acta Mechanica Sinica/Lixue Xuebao 29(4):550–556
14. El Said B, Hallett SR (2018) Multiscale surrogate modelling of the elastic response of thick composite structures with embedded defects and features. Compos Struct 200:781–798
15. Haeri A, Fadaee MJ (2016) Efficient reliability analysis of laminated composites using advanced kriging surrogate model. Compos Struct 149:26–32
16. Díaz J, Cid Montoya M, Hernández S (2016) Efficient methodologies for reliability-based design optimization of composite panels. Adv Eng Softw 93:9–21
17. Omairey SL, Dunning PD, Sriramula S (2019) Multiscale surrogate-based framework for reliability analysis of unidirectional FRP composites. Compos Part B: Eng 173:106925
18. Omairey SL, Dunning PD, Sriramula S (2020) Multi-scale reliability-based design optimisation framework for fibre-reinforced composite laminates. Eng Comput **ahead-of-print** (-)
19. Riaño L, Belec L, Chailan J, Joliff Y (2018) Effect of interphase region on the elastic behavior of unidirectional glass-fiber/epoxy composites. Compos Struct 198:109–116
20. Omairey SL, Dunning PD, Sriramula S (2019) Development of an ABAQUS plugin tool for periodic RVE homogenisation. Eng Comput 35(2):567–577
21. Kassapoglou C (2010) Design and analysis of composite structures: with applications to aerospace structures
22. Reddy JN (2004) Mechanics of laminated composite plates and shells: theory and analysis. Ringgold Inc, Portland, USA
23. Jones RM (1999) Mechanics of composite materials. Brunner-Routledge, New York, London
24. Xia Z, Zhang Y, Ellyin F (2003) A unified periodical boundary conditions for representative volume elements of composites and applications. Int J Solids Struct 40(8):1907–1921
25. Sakata S, Ashida F, Enya K (2012) A Microscopic failure probability analysis of a unidirectional fiber reinforced composite material via a multiscale stochastic stress analysis for a microscopic random variation of an elastic property. Comput Mater Sci 62:35–46
26. Henrik ST, Branner K, Leon Mishnaevsky J, John DS (2013) Uncertainty modelling and code calibration for composite materials. J Compos Mater 47(14):1729–1747
27. Geers MGD, Kouznetsova VG, Brekelmans WAM (2010) Multi-scale computational homogenization: Trends and challenges. J Comput Appl Math 234(7):2175–2182
28. Hahn GJ (1972) Sample sizes for monte carlo simulation. IEEE Trans Syst Man Cybern **SMC-2**(5): 678–680

Understanding the Binding Affinity and Specificity of miRNAs: A Molecular Dynamics Study

Swarnima Kushwaha, Ayushi Mandloi, and Shibasish Chowdhury

Abstract MicroRNAs (miRNA) are endogenously produced small (21–27 nucleotides long) non-coding RNA molecules that can play post-transcriptional regulation of gene expression either by cleavage of the mRNA strand or by translational repression. However, recognizing mRNA targets by miRNAs is not clearly understood. It is observed that one miRNA molecule can target hundreds of different mRNA sites and different miRNAs can target a single mRNA site. The majority of miRNA target prediction algorithms mainly consider the complementarity on the seed region of miRNA for their target selection. To explore the role of different regions of miRNA for target selection, explicit solvent unrestrained molecular dynamics simulation has been performed on miRNA-RNA duplex. Our studies revealed that seed region complementary base pairing is sufficient to maintain the integrity of the duplex, whereas bulge and 3′ end regions are highly flexible which can interact with target mRNA through non-canonical hydrogen bonds and regulate translational repression.

Keywords miRNA · RNA duplex structure · Molecular dynamics simulation · Hydrogen bonds

1 Introduction

Small non-coding RNAs, called microRNAs (miRNAs), are 21–27 nucleotides long that carry out and regulate various biological processes. Over the years, researchers have found this small, non-coding RNAs have crucial role in eukaryotic gene expression by their ability to degrade an mRNA or suppress translation [1]. Primary miRNAs are generated from genes which are then cleaved by the microprocessor complex giving rise to precursor-miRNAs. These pre-miRNAs are then interacted with endoribonuclease DICER to form mature miRNAs [2]. Furthermore, these mature miRNAs

S. Kushwaha · A. Mandloi · S. Chowdhury (✉)
Department of Biological Sciences, Birla Institute of Technology and Sciences, Pilani Campus, Pilani, RJ 333031, India
e-mail: shiba@pilani.bits-pilani.ac.in

© The Author(s), under exclusive license to Springer Nature Singapore Pte Ltd. 2022   229
R. Srinivas et al. (eds.), *Advances in Computational Modeling and Simulation*,
Lecture Notes in Mechanical Engineering,
https://doi.org/10.1007/978-981-16-7857-8_19

act on mRNA to degrade or silence their expression with the help of an RNA-induced silencing complex (RISC). The potent role of target recognition is played by the 5' end of miRNA, also known as the "seed" region. This core element consists of 2–7 nucleotides and forms central components of many computer-based miRNA target prediction tools. Target selection in RNA silencing is governed by this "seed sequence" at the 5' end of the guide strand, with overall binding free energy (ΔG) values of this miRNA-mRNA interaction. Though the bulge (mid) and 3' region of the miRNA is inconsequential it can attune this activity in certain unforeseen circumstances. Currently, a large number of miRNA target prediction tools like miRanda, TargetScan, DIANA-microT, MirTarget2, SVMicrO, PITA, and RNAhybrid are available, which majorly utilizes information regarding seed pairing, evolutionary conservation of miRNA sequences among different species, the free energy of miRNA target binding, and miRNA site accessibility to predict mRNA targets [3]. However, various studies also indicated that the interaction of miRNA with target mRNA might not be the sole determining factor for gene regulation as miRNA localization, target mRNA concentration, and many protein factors associated with RISC are involved in this process [4]. Interestingly, it was shown that mismatches within the seed region increase the specificity and efficiency of miRNA-guided silencing by extensive complementarity on the non-seed regions [5, 6]. In this regard, G:U wobble pairing may play a crucial role in target selection [7].

Expression levels of both the mRNA and the miRNA and their probable matching binding sites on other mRNAs need to be considered to understand the endogenous regulation of mRNA by the miRNA. Various experimental studies, including comparative genomics and high-throughput experimental studies, indicated that a miRNA binds to hundreds of sites on mRNA and a particular gene can be regulated by hundreds of miRNAs [8]. It has also been noticed that most of the predicted miRNA targets undergo changes at the mRNA and protein levels when the miRNA expression is disturbed. Similarly, different miRNAs have an unpredictable effect on target gene expression. There are approximately 200–255 miRNAs in the human genome [9], and in terms of their expression levels, some miRNAs are expressed more than 1000 copies per cell [10]. Although miRNA-regulated gene expression is now well established, exact identification of miRNA target is still elusive as miRNA can bind multiple sites within an mRNA target that cannot be easily studied experimentally [11–13]. Hence, computational methods including molecular dynamics (MD) simulation for miRNA target prediction become critical. Due to the dynamics and polymorphic nature of the miRNA target complex, MD simulation becomes very useful in accurately predicting the stability of the miRNA target complex with precise, detailed knowledge of interaction [14].

hsa-miR-7-5p is one of the widely studied miRNAs with various roles in multiple cancers, including negative regulation of sprouting angiogenesis and the inhibition of the EGF receptor and the Akt pathway [15, 16]. It also inhibits colorectal cancer cell proliferation and induces apoptosis by targeting XRCC2 [17]. A miRTarBase search has resulted in a total of 579 mRNA targets of hsa-miR-7-5p [18]. To understand the molecular basis of target selection of miRNA, we have selected two mRNA targets of hsa-miR-7-5p within PAK1 (serine/threonine-protein kinase) gene. This study

Fig. 1 Schematic diagram representing potential hydrogen bonding scheme and basepair interaction scheme of has-miR-7-5p (miRNA) with (**a**) mRNA target 1 (**b**) mRNA target 2. The bases within the seed region (2–8 nucleotides of miRNA) and bulge region (9–12) are shown in pink and green colors, respectively. The canonical Watson–Crick (WC) hydrogen bond scheme was shown by solid line within the base pair, while a possible hydrogen bonding scheme within G:U wobble pair is shown in the dotted line

evaluated the stability and interaction pattern of hsa-miR-7-5p with its two mRNA targets using MD simulation. The role of a different region of miRNA (seed, bulge, and 3′ end) on mRNA recognition is also explored through MD simulation.

2   Materials and Methods

2.1   Initial Duplex Model

The initial model of has-miR-7-5p (5′-UGGAAGACUAGUGAUUUGUUGUU-3′) and its both targets mRNA (5′-UAAAUAAAUGUUUCUAGUCUUCCG-3′ and 5′-UUUAU AACAUUGAGAGGUUUUCUA-3′) were built using RNAComposer server [19]. The starting A-RNA duplex models of miRNA with both the mRNA targets were modeled using the experimentally verified interaction pattern [18]. The hydrogen bonding scheme and basepair interaction scheme of has-miR-7-5p with both the targets are shown in Fig. 1a, b. The initial model of miRNA-target-1 duplex consists of a total of 51 hydrogen bonds between two strands, whereas, initially, 56 hydrogen bonds are observed for miRNA-target-2 duplex. In the subsequent discussion, the miRNA-target-1 duplex will be termed as target 1, while the miRNA-target-2 duplex will be designated as target 2.

2.2   Molecular Dynamics Simulation

The structural fluctuations and stability of the miRNA-mRNA complex were analyzed by time-dependent molecular dynamics (MD) simulation studies. Each miRNA-mRNA complex was simulated in three steps—generation of the simulation environment, equilibration phase, and unrestrained production simulation. The tleap module of AMBER 18 [20] was used to neutralize the charge of the initial miRNA-mRNA A-RNA duplex model by placing a total of 46 Na^+ ions at positions with high electronegative potential. miRNA-mRNA complex and counterions were then placed in a pre-equilibrated octahedron box of TIP3P water molecules. The periodic box of water was extended to a distance of 10 Å from RNA duplex and counterions. The prepared system was minimized in two stages. In the first stage, solvent and ions were subjected to 5000 steps (1000 cycles of steepest descent followed by 4000 cycles of conjugate gradient minimization) of minimization while RNA duplex was constrained. In the next stage, entire system was subjected to 5000 cycles of unconstrained minimization. Molecular dynamics was performed with the SANDER module of the AMBER 18 program using the all-atom RNA ff99OL3 force field [21]. The particle mesh Ewald method (PME) was used for the calculation of electrostatic interactions. Periodic boundary conditions were imposed in all directions. The long-range electrostatic interactions have been calculated without any truncation, while a 12 Å cutoff was applied to Lennard–Jones interactions. SHAKE algorithm [22] was applied to constrain the bond involving hydrogens. The temperature was controlled at 300 K using Langevin dynamics with the collision frequency 1. A time step of 2 fs was used and the structures were saved at every 10 ps interval for the entire duration of the MD run. Equilibrium simulation was comprised of the solute's restrained heating phase (temperature was increased from 0 to 300 K within 50 ps) of constant NVT simulation with restraints on the RNA atoms, constant NPT simulation (150 ps) with restraints on RNA atoms, and unrestrained simulation (300 ps). The unrestrained 250 ns simulation with NPT ensembles at 300 K was considered as production simulation.

2.3 MD Trajectory and Structural Analysis

MD trajectories were analyzed by using the PTRAJ module of AMBER 18. The MD average structures are initially obtained from the coordinates saved between 290 and 300 ns. These structures were first solvated and charges are neutralized with TIP3P water molecules and Na^+ ions, respectively. The solvated structures were then subjected to 10,000 cycles of unrestraint energy minimization. The energy minimized MD average structures were considered as MD average structures. RNA duplex structures were visualized in VMD [23] and structural parameters were calculated using the program NUPARM [24]. The binding free energy for the association of miRNAs with its target was calculated using the MMPBSA module of AMBER 18 [25].

3   Results and Discussion

3.1   *Stability of miRNA-mRNA Interaction*

To study the interaction, specificity, and structural stability of has-miR-7-5p molecule with its mRNA (PAK1) targets, we performed unrestrained 300 ns explicit solvent molecular dynamics simulations of has-miR-7-5p with its two mRNA targets (target 1 and target 2). The temperature, density, and potential energy plots (Fig. 2) show that the simulated systems are equilibrated well within 500 ps.

The root mean square (RMS) deviation values for the initial energy minimized structure was calculated for all the heavy atoms in both the RNA duplex and for the seed, bulge, and 3′ end region separately (shown in Fig. 3).

It is observed that during the simulation, the average RMS deviation of target 1 (5.88 Å) from its initial model as well as RMSD fluctuation of target 1 complex (standard deviation of 1.41) is more than that of target 2 (5.77 Å and 1.28, respectively), which is mainly due to the structural alteration in the 3′ end of the target 1 complex. However, the average RMS deviation in the bulge region of target 2 (3.63 Å) is significantly higher than that of target 1 (1.48 Å). RMSD plot (Fig. 3) suggests that 3′ end of target 1 was altered considerably from its initial model during the simulation, whereas the bulge region of target 2 was changed significantly from its initial model. The structural fluctuation and deviation from the initial models are marginal for the

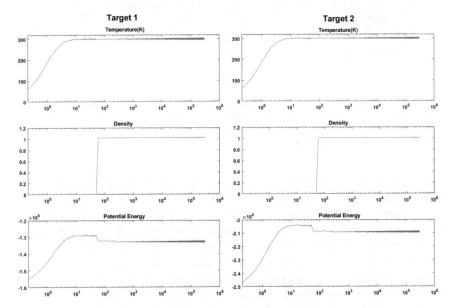

Fig. 2 System temperature (in Kelvin), density (g/ml), and total potential energy (in kcal/mol) are shown during the entire 300 ns of MD simulation of target 1 (first column) and target 2 (second column). The logarithm scale of time (in picosecond) is plotted along the x-axis

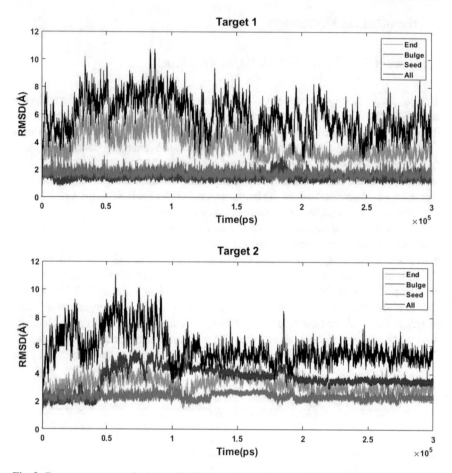

Fig. 3 Root mean square deviation (RMSD) profiles during the MD simulation of target 1 and target 2 with respect to their respective initial energy minimized structures. The four plots in each figure correspond to the RMSD for all atoms (black lines), the seed region (pink lines), bulge region (blue lines) and 3′ end (green lines) region

seed regions of both the targets as perfect complementarity was present within seed regions. Nevertheless, the compactness of both RNA duplex was retained during the entire simulation as the radius of gyration of both RNA duplexes remains close to their initial values (Fig. 4), indicating the robustness of the miRNA-mRNA duplex.

The superimposition of initial energy minimized miRNA-mRNA duplex models and MD average structures (Fig. 5a, b) during the last 10 ns of dynamics also confirms the result obtained from RMS analysis. It is also noticed that the RNA backbone of 3′ end of MD average target 1 structure is less smooth than that of the remaining duplex indicating that 3′ end of the duplex molecule undergoes larger fluctuation during the simulation and it is possibly more flexible (Fig. 5a).

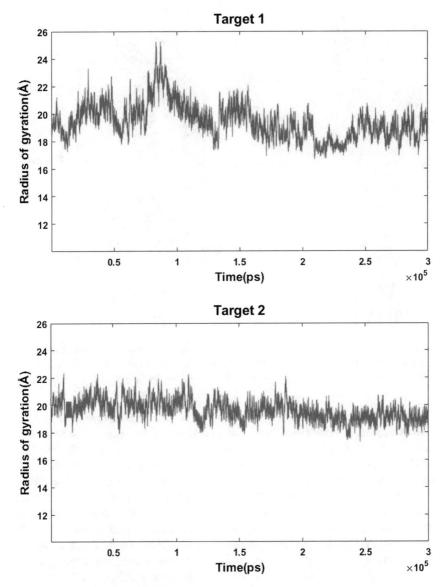

Fig. 4 Trajectories showing the radius of gyration of target 1 and target 2. The average radius of gyration of both the targets during the last 10 ns of the simulation was very close to each other and was 18.9 Å and 19.0 Å, respectively

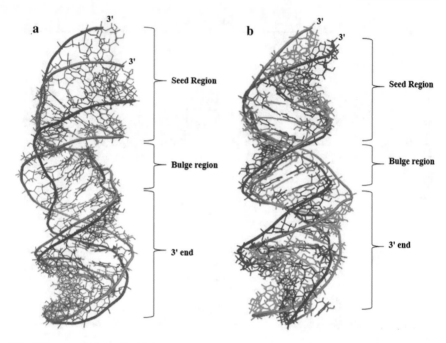

Fig. 5 Energy minimized initial RNA duplex (in blue lines) is shown superimposed on energy minimized MD average structures (in red lines) of (**a**) target 1 and (**b**) target 2. The RNA backbone is shown by tube. The 3′ end of mRNA target is labeled in both the figures

On the other hand, the bulge region of target 2 is more flexible than other parts of the RNA duplex. It is observed that the RNA backbone of the bulge region of the MD average structure significantly deviated from that of the initial model (Fig. 5b). The MMPBSA-based binding free energy of both the miRNA-mRNA complexes was calculated during MD simulation using the snapshots saved between 100 and 300 ns of simulation. The binding free energy of target 1 was − 15.56 kcal/mol, which is almost similar to the average binding free energy of 171 experimentally tested miRNA-mRNA interactions in *Drosophila melanogaster* (−15.6 kcal/mol) [26]. In comparison, the binding free energy of target 2 was 4.33 kcal/mol indicating that the interaction of miRNA with target 1 possibly produces effective gene silencing as the complex of miRNA with target 1 is about 20 kcal/mol more stable than that of target 2.

3.2 Structural Parameters of RNA Duplex

The RNA backbone geometry of initial energy minimized and MD averaged duplex of both the targets are analyzed and tabulated in Table 1.

Table 1 Backbone torsion angles, the glycosidic torsion angle χ, and the ribose sugar ring pseudorotation phase angle P values for the A-RNA fiber model [27] and initial as well as MD average structure of both target 1 and target 2 are tabulated. The values are averaged over all residues in a duplex. The standard deviation values are given within parentheses. All values are in degree (°)

–	A-RNA	Target 1		Target 2	
–	–	Initial	MD average	Initial	MD average
α	−62	−98 (39)	−85 (31)	−100 (44)	− 107 (63)
β	180	171 (20)	174 (8)	167 (33)	173 (8)
γ	47	79 (34)	69 (17)	88 (47)	65 (9)
δ	83	89 (16)	82 (10)	87 (16)	87 (21)
ε	−151	− 155 (26)	−160 (14)	− 161 (26)	− 161 (25)
ζ	−74	−83 (53)	− 73 (35)	−92 (70)	− 107 (88)
χ	−166	−152 (29)	− 154 (23)	− 144 (20)	−150 (27)
P	13	27 (44)	26 (24)	21 (41)	28 (53)

It is observed that backbone conformation of majority of nucleotides in MD averaged target 1 adopted canonical A-RNA conformation [27] with alpha (α), beta (β), gamma (γ), delta (δ), epsilon (ε), and zeta (ζ) values are in *gauche⁻* (*g⁻*), *trans* (*t*), *gauche⁺* (*g⁺*), *gauche⁺* (*g⁺*) *trans* (*t*), and *gauche⁻* (*g⁻*) region, respectively. Only ζ torsion angle of the first nucleotide of RNA target and α and γ torsion angles of the second residue of RNA target are in *g⁺*, *g⁺*, and *t* region, respectively. The backbone geometry of target 1 becomes more uniform during the MD simulation as the standard deviation of torsion angles in the MD average structure is much smaller than that of the initial model (Table 1). The glycosidic torsion angles of MD average structure of target 1 remain in *anti*-orientation while the pseudorotation phase angles (P) are around C3′-endo region. During the MD simulation, the RNA overall backbone geometry of target 2 duplex also remains in canonical A-RNA conformation [27]. However, in comparison to target 1, the backbone of target 2 is less uniform as we observed that quite a few torsion angles adopted non-canonical A-RNA conformation. For example, few α torsion angles are in the *g⁺* or *t* region, δ torsion angles are in the *t* region, and ζs are in the *g⁺* region. Few ribose sugars are moved from canonical C3′-endo sugar conformation to C4′-exo conformation.

The inter- and intra-basepair structural parameters were calculated for the MD average structures of both the targets, and then compared with initial models and fiber A-RNA duplex [27]. It is noticed that the seed region of both the targets maintained the A-RNA helical geometry during the MD simulation and average local helical structural parameters are very close to the fiber RNA model (Table 2).

However, in comparison to target 1, the variation of structural parameters like "Inclination," "Tip," "Twist," "dx," and "dz" is larger in target 2 (as seen from standard deviation values in Table 2), indicating that seed region of target 1 possibly more firmly held through hydrogen bonds. A similar trend is also observed in the case of intra-basepair parameters like "Propeller twist," "Buckle," and "Opening." Local step parameters within the seed regions are also close to the fiber RNA structure.

Table 2 Local helical, intra-basepair, and local step parameters of A-RNA fiber model, MD average target 1, and MD average target 2 structure are shown. The parameters are calculated separately for seed, bulge, and 3′ end region of MD averaged duplex structure. The standard deviation values are given within parentheses. The rotational parameters are given in degrees (°) and translational parameters are given in Å

	A-RNA	Target 1			Target 2		
		Seed	Bulge	3′end	Seed	Bulge	3′ end
Inclination	15.8	14.5(10.2)	23.3(15.3)	19.1(22.2)	22.7(19.1)	10.6(34.3)	16.1(16.1)
Tip	0	−0.2(4.4)	6.4(9.5)	2.4(11.1)	8.7(6.8)	11.8(40.1)	−1.8(3.4)
Twist	31.5	30.4(3.6)	32.9(7.5)	36.9(12.3)	32.5(5.6)	−13.2(109)	32.7(8.8)
X-displacement	−5.3	−3.8(2.3)	−8.0(6.4)	−4.4(3.0)	−5.1(4.0)	−4.2(6.7)	−5.1(3.3)
Y-displacement	0	0.1(1.9)	0.2(2.5)	−1.0(2.9)	0.8(1.3)	1.0(2.6)	−5.1(3.3)
Z-displacement	2.8	3.0(0.6)	2.3(1.0)	2.8(0.6)	2.5(0.9)	2.3(1.3)	2.8(0.5)
Buckle	0	2.8(8.5)	9.2(10.3)	−1.8(10.4)	−5.9(16.5)	0.2(14.0)	2.3(4.8)
Propeller Twist	14.5	−16.7(10.2)	−19.1(12.4)	−23.1(10.4)	−23.1(13.6)	−12.5(13.1)	−19.3(7.8)
Opening	−4.2	18.4(4.7)	17.8(11.8)	15.3(4.0)	20.8(6.8)	16.9(4.7)	13.9(7.6)
Tilt	0	0.2(2.7)	−3.1(3.9)	−0.3(7.5)	4.7(3.5)	4.9(8.7)	0.8(1.5)
Roll	8.9	7.6(5.6)	14.2(12.0)	8.1(7.1)	12.2(10.7)	15.2(7.1)	7.7(8.5)
Twist	31.5	28.9(3.6)	27.3(8.1)	34.7(12.8)	27.8(6.3)	27.2(13.1)	30.6(9.7)
Shift	0	−0.1(0.9)	0.0(1.0)	0.7(1.8)	−0.7(0.4)	−0.9(1.5)	0.3(0.5)
Slide	−2.1	−1.0(0.7)	−1.2(0.8)	−1.5(1.1)	−1.0(0.6)	−1.1(0.2)	−1.5(0.7)
Rise	3.52	3.4(0.2)	3.5(0.2)	3.3(0.6)	3.4(0.3)	3.4(0.2)	3.3(0.2)

The "Slide" value is slightly smaller in the fiber model than that of the MD average structures. Since ample mismatches are present within the bulge and 3′ end region of both the targets, local helical, intra-basepair, and local step parameters within those regions have substantially deviated from structural parameters of canonical A-RNA fiber structure. However, when compared with target 1, deviation from the fiber model is more in the bulge region of target 2, whereas 3′ end of target 1 is more deviated from 3′ end region of target 2 (Table 2). This observation reinforced that 3′ end region of target 1 and bulge region of target 2 are more flexible.

3.3 Hydrogen Bonding Pattern

The hydrogen bonding pattern and its flexibility were monitored throughout the MD simulation. It is noticed that for both the targets, all the canonical Watson–Crick hydrogen bonds within the seed region were retained during the entire length of simulation to provide the required integrity of the RNA duplex. Interestingly, in the case of target 2, even the less favorable G:U pairing within the seed region retains its hydrogen bonds with occasional deformation (Fig. 6a, b).

However, alteration in hydrogen bond patterns is observed in both bulge and 3′ end region of both the duplexes. In the case of target 1, all the bases within the bulge region were hydrogen bonded. Interestingly, a U:U base pair was stabilized by two hydrogen bonds (Fig. 7a). However, during the simulation, a translational movement of uracil bases altered the hydrogen bonding patterns within the U:U pair (Fig. 7b). This alteration of hydrogen bonding distance was monitored during the simulation, which is shown in Fig. 7c.

On the contrary, we did not observe any stable hydrogen-bonded pairing between RNA bases in the bulge region of target 2. During the simulation, few transient hydrogen bonds are formed among few bases, making the bulge region extremely flexible. Despite having a larger number of canonical A:U base pairs (with 6 A:U basepair in total) within the 3′ region, in comparison with target 2, target 1 suffered larger deformation during simulation, which is mainly because of the formation of a larger number of G:U wobble pairs within the 3′-end of target 2. Few of these G:U pairs were even stabilized by two hydrogen bonds (Fig. 8).

Trajectory analysis has revealed that, on average, 39.6 hydrogen bonds were present within target 1, whereas 40 hydrogen bonds were observed within target 2, indicating that the number of hydrogen bonds in these structures may not be a deciding factor for their relative stability.

4 Conclusion

Present MD simulations explore the interaction patterns of miRNA-mRNA complexes, which are crucial for their stability. Our simulation data indicated that

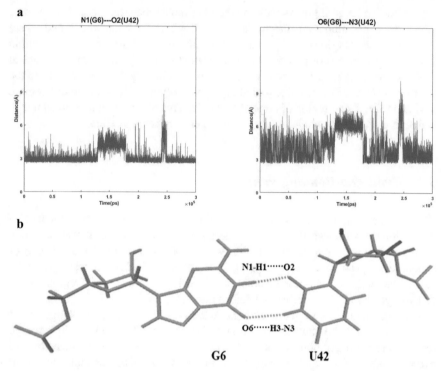

Fig. 6 (**a**) Trajectories showing the distances between N1 (G6) and O2 (U42) and O6 (G6) and N3 (U42) in target 2. Time (in picosecond) is along the x-axis and distance (in Å) is along the y-axis. (**b**) Hydrogen bonding scheme of G6:U42 pair in energy minimized MD average structure of target 2 is shown. The dotted line indicates hydrogen bonding between N1-H1–O2 and O6–H3-N3

the bulge (mid) and 3′ end regions of the duplex are more flexible. Non-canonical hydrogen bonds, including G:U wobble pairs, are frequently formed within these regions, whereas seed regions are relatively rigid with stable hydrogen-bonded base pairing. G:U pairings are found to be critical to the stability of the duplex. Non-seed parts are possibly playing a critical role in alternative target selection.

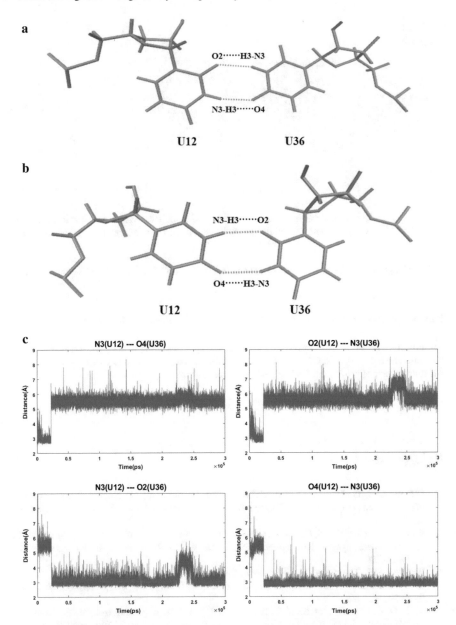

Fig. 7 Hydrogen bonding scheme of U12:U36 pair in (**a**) initial energy minimized target 1. The dotted line indicates hydrogen bonding between O2(U12)—H3-N3(U36) and N3-H3(U12)—O4(U36) (**b**) energy minimized MD average structure of target 1 is shown. The dotted line indicates hydrogen bonding between N3-H3(U12) —O2(U36) and O4(U12) —H3-N3(U36) (**c**) Trajectories showing the distances between N3 (U12) and O4 (U36), O2 (U12) and N3 (U36), N3 (U12) and O2 (36), O4 (U12) and N3 (U36) in target 1. Time (in picosecond) is along the x-axis and distance (in Å) is along the y-axis

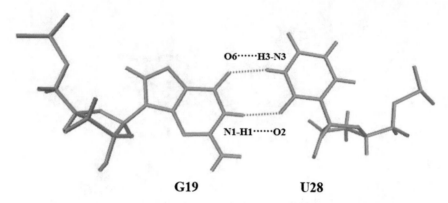

Fig. 8 Hydrogen bonding scheme of a representative G:U wobble base pairing is shown for energy minimized MD average target 2 structure. Dotted lines indicate the hydrogen bond between atoms

Acknowledgements We would like to acknowledge the HPC facilities of BITS-Pilani, Pilani campus.

Funding Sources This research did not receive any specific grant from funding agencies.

Author Contributions SC conceived, designed, and supervised the overall project; AM carried out initial modeling work; and SK, AM, and SC performed data analysis and wrote the manuscript. All authors have approved the final version of the manuscript.

Conflict of Interest All the authors declare no conflict of interest.

References

1. O'Brien J, Hayder H, Zayed Y, Peng C (2018) Overview of microRNA biogenesis, mechanisms of actions, and circulation. Front Endocrinol (Lausanne) 9:1–12. https://doi.org/10.3389/fendo.2018.00402
2. Winter J, Jung S, Keller S, Gregory RI, Diederichs S (2009) Many roads to maturity: MicroRNA biogenesis pathways and their regulation. Nat Cell Biol 11(3):228–234. https://doi.org/10.1038/ncb0309-228
3. Peterson SM, Thompson JA, Ufkin ML, Sathyanarayana P, Liaw L, Congdon CB (2014) Common features of microRNA target prediction tools. Front Genet 5. https://doi.org/10.3389/fgene.2014.00023
4. Riolo G, Cantara S, Marzocchi C, Ricci C (2021) miRNA targets: from prediction tools to experimental validation. Methods Protoc 4(1):1–20. https://doi.org/10.3390/mps4010001
5. Brancati G, Großhans H (2018) An interplay of miRNA abundance and target site architecture determines miRNA activity and specificity. Nucleic Acids Res 46(7):3259–3269. https://doi.org/10.1093/nar/gky201
6. Ameres SL, Martinez J, Schroeder R (2007) Molecular basis for target RNA recognition and cleavage by human RISC. Cell 130(1):101–112. https://doi.org/10.1016/j.cell.2007.04.037
7. Broughton JP, Lovci MT, Huang JL, Yeo GW, Pasquinelli AE (2016) Pairing beyond the seed supports MicroRNA targeting specificity. Mol Cell 64(2):320–333. https://doi.org/10.1016/j.molcel.2016.09.004

8. Hausser J, Zavolan M (2014) Identification and consequences of miRNA-target interactions-beyond repression of gene expression. Nat Rev Genet 15(9):599–612. https://doi.org/10.1038/nrg3765

9. Lai EC (2003) MicroRNAs: runts of the genome assert themselves. Curr Biol 13(23):925–936. https://doi.org/10.1016/j.cub.2003.11.017

10. Alvarez Garcia I, Miska EA (2007) The microRNAs of C elegans. MicroRNAs from basic. Sci Dis Biol 7–21. https://doi.org/10.1017/CBO9780511541766.004

11. Villa A, Wöhnert J, Stock G (2009) Molecular dynamics simulation study of the binding of purine bases to the aptamer domain of the guanine sensing riboswitch. Nucleic Acids Res 37(14):4774–4786. https://doi.org/10.1093/nar/gkp486

12. Wang Y, Li Y, Ma Z, Yang W, Ai C (2010) Mechanism of microRNA-target interaction: Molecular dynamics simulations and thermodynamics analysis. PLoS Comput Biol 6(7):5. https://doi.org/10.1371/journal.pcbi.1000866

13. Minchington TG, Griffiths-Jones S, Papalopulu N (2020) Dynamical gene regulatory networks are tuned by transcriptional autoregulation with microRNA feedback. Sci Rep 10(1):1–13. https://doi.org/10.1038/s41598-020-69791-5

14. Bochicchio A, Krepl M, Yang F, Varani G, Sponer J, Carloni P (2018) Molecular basis for the increased affinity of an RNA recognition motif with re-engineered specificity: a molecular dynamics and enhanced sampling simulations study. PLoS Comput Biol 14(12):1–27. https://doi.org/10.1371/journal.pcbi.1006642

15. Yue K, Wang X, Wu Y, Zhou X, He Q, Duan Y (2016) MicroRNA-7 regulates cell growth, migration and invasion via direct targeting of PAK1 in thyroid cancer. Mol Med Rep 14(3):2127–2134. https://doi.org/10.3892/mmr.2016.5477

16. Gajda E, Grzanka M, Godlewska M, Gawel D (2021) The role of miRNA-7 in the biology of cancer and modulation of drug resistance. Pharmaceuticals 14(2):1–24. https://doi.org/10.3390/ph14020149

17. Saydam et al O (2012) NIH Public Access, vol 71, no 3, pp 852–861 https://doi.org/10.1158/0008-5472.CAN-10-1219.miRNA-7

18. Huang HY et al (2020) MiRTarBase 2020: updates to the experimentally validated microRNA-target interaction database. Nucleic Acids Res 48(D1):D148–D154. https://doi.org/10.1093/nar/gkz896

19. Popenda M et al (2012) Automated 3D structure composition for large RNAs. Nucleic Acids Res 40(14):1–12. https://doi.org/10.1093/nar/gks339

20. Case DA et al (2018) Amber 2018. The University of California, San Francisco 2018:1–923

21. Zgarbová M et al (2011) Refinement of the Cornell et al. Nucleic acids force field based on reference quantum chemical calculations of glycosidic torsion profiles. J Chem Theory Comput 7(9):2886–2902. https://doi.org/10.1021/ct200162x

22. Ryckaert JP, Ciccotti G, Berendsen HJC (1977) Numerical integration of the cartesian equations of motion of a system with constraints: molecular dynamics of n-alkanes. J Comput Phys 23(3):327–341. https://doi.org/10.1016/0021-9991(77)90098-5

23. Humphrey W, Dalke A, Schulten K (1996) VMD: visual molecular dynamics. J Mol Graph 14:33–38. https://www.tapbiosystems.com/tap/products/index.htm

24. Bansal M, Bhattacharyya D, Ravi B (1995) Acid Structures. Cabios 11(3):281–287

25. Miller BR, Mcgee TD, Swails JM, Homeyer N, Gohlke H, Roitberg AE (2012) MMPBSA.py : an efficient program for end-state free energy calculations

26. Ragan C, Zuker M, Ragan MA (2011) Quantitative prediction of miRNA-mRNA interaction based on equilibrium concentrations. PLoS Comput Biol 7(2). https://doi.org/10.1371/journal.pcbi.1001090.

27. Arnott S, Hukins DWL, Dover SD, Fuller W, Hodgson AR (1973) Structures of synthetic polynucleotides in the A-RNA and A'-RNA conformations: X-ray diffraction analyses of the molecular conformations of polyadenylic acid polyuridylic acid and polyinosinic acid polycytidylic acid. J Mol Biol 81(2):107–122. https://doi.org/10.1016/0022-2836(73)90183-6

Printed in the United States
by Baker & Taylor Publisher Services